# The Papers of Independent Authors

I0486277

## Volume 7

Israel    2008

Design – Gelfand L.M.
Technical editing – Gelfand L.M.

It is sent in a press on March, 16th 2008
It is printed in the USA, Lulu Inc., catalogue **2211542**
ISBN 978-1-4357-1643-8
Site with information for the author - http://izdatelstwo.com
The contact information - publisher-dna@hotmail.com
Fax: ++972-8-8691348
Address: POB 15302, Bene-Ayish, Israel, 79845
The form of the reference:
    Author. Paper. «The Papers of Independent Authors», Publisher «DNA», Israel, Printed in USA, Lulu Inc., catalogue 2211542, vol. 7, 2008.

Truth - the daughter of time, instead of authority
Frensis Bacon

Everyone has the right to freedom of opinion and expression; this right includes freedom to hold opinions without interference and to seek, receive and impart information and ideas through any media and regardless of frontiers.
United Nations OrganizationUniversal. Declaration of Human Rights. Article 19.

### From the Publisher

"The Papers of Independent Authors" - versatile scientific and technical printed magazine. The magazine accepts articles to the publication from all countries. The following rules are complied by this:

- Articles are not reviewed and the publishing house is not responsible for a content and style of publications;
- The magazine is registered in the international qualifier of books ISBN, transferred and registered in national libraries of Russia, USA, Israel;
- The priority and copyrights of the author of paper are provided with registration of magazine in ISBN;
- The commercial rights of the author of paper are kept for the author;
- The magazine is published in Israel and printed in the USA;
- The printed magazine is on sale under the cost price (for publishing house), and in an electronic kind provides free of charge;
- The author pays for the publication.

This magazine - for those authors who are assured of himselves and do not require approval of the reviewer. Us often reproach that papers are not reviewed. But the institute of reviewing is not the ideal filter – it passes unsuccessful papers and detains original works. Without analyzing the numerous reasons of it, we shall notice only, that if the reader outstanding ideas can remain unknown persons can filter bad papers. Therefore we wish to give to scientists and to engineers the right (similarly to writers and artists) to be published without reviewing and to not to spend years for the statement of the ideas.

*Solomon I. Khmelnik.*

# Contents

# Series: **Algorithms and Programming**

**Alexander M. Gelfand, Solomon I. Khmelnik**

# Discrete filtration of Multivariate Correlated Nonstationary Processes

## Contents

## Summary

We consider a vector stochastic process with stationary increments of a predetermined order, whose components are linearly dependent, i.e. in the absence of noise vector process components are constrained by a system of linear equations (constraints). The interdependence of stochastic processes can be determined by a static or a dynamic model. The constraints can be maintained rigidly or with a specified error. We offer a method allowing in these conditions synthesis of an optimum filter structure. This method works in cases where no information about signal and noise static properties is available.

# 1. Introduction

The works [1, 2, 3, 4] review various problems of synthesis of an optimum discrete filter for random series with stationary increments of a predetermined order (RSSI) for brevity's sake hereinafter referred to as *RSSI filters*. Below, these problems will be generalized, supplemented and classified. For this purpose only end results needed by a developer of a control system containing filters.

For a situation characterized by nonstationarity of measurements and the static model of a control object, the problem was formerly considered on the basis of an empirically introduced filter structure, and its solution was reduced to finding filter parameter values. The method proposed allows to *synthesize a* **structure** *of an optimum filter*. The method works even when no information about signal and noise static properties is available. It is specifically found that an exponential smoothing filter is a special case of the proposed filter and is optimal according to the criterion laid down in this work. We do not know of analogues to the algorithm proposed. The method covers *interdependent* RSSI's. The interdependence between them yields some additional information that improves the quality of filtration. If we know the filter structure, we can filter such processes in real-time systems.

For *dynamic systems* Kalman filter is widely used. As compared with it, the filter proposed

- has the capability of finding filter parameters in the absense of information necessary for Kalman filter synthesis (knowledge of auto- and intercorrelation functions of signals and noise),
- requires considerably less amount of on-line calculations owing to the less complex structure of the synthesized filter.

Furthermore, there is no solution to the problem in the mentioned formulation using Kalman filtration theory.

The synthesis problem can be formulated in the following way We consider a vector stochastic process with stationary increments of the $p$-th order whose components are linearly dependent, , i.e. in the absence of noise vector process components are constrained by a system of linear equations (constraints). This system can be either static or dynamic. The constraints can be maintained rigidly or with a specified error. We need to synthesize a filter in such a way that this dependence would remain for signal estimates. Various versions of this problem have been solved [1, 2, 3, 4] by methods that are used for control design synthesis [5].

The meaning of this formulation is as follows. Let us assume that there is a time function $Z(t)$, whose $p$-th derivative has the small value of $\Delta(t) \approx 0$. The filter is constructed in such a way that in the process of noise filtration of function $Z(t)$, i.e. during the calculation of function $L(t)$ estimate, the value of $(1 - \beta) \cdot \Delta(t) + \beta \cdot [Z(t) - L(t)]$ would get minimized, where $\beta$ is a weight coefficient and $0 \le \beta \le 1$.

## 2. Multiply correlated vector stochastic processes

Below we consider a vector RSSI-$p$ with stationary increments of the $p$-th order whose components are linearly dependent, i.e. in the absense of noise the system of linear equations comprising a control object model is given by $Model\ (Z) = 0$. E.g., it may be of the form $A \cdot Z = C$, where $A$ is a known matrix, and $C$ is a known vector. The constraints can be maintained with different degrees of strictness (including strong constraints). We need to synthesize a filter in such a way that that this dependence would remain for signal estimates. At every instant $j$ for each $i$-th process an additive mixture $Z_i(j)$ of useful signal and noise is observed. This mixture will be the input of the filter whose output we shall call $L_j$. As a result of the synthesis a time **independent** matrix $B$ is formed. With a known matrix $B$, the filtration, i.e. the calculation of optimum estimate $L_i(j)$, is performed by formula $L(j) = B \cdot W(j)$, where $W(j)$ is a determined function of measurements vector $Z(j)$ and values of vectors $Z(k)$ and $L(k)$ at past instants of time.

So the mathematical model of a control object $Model\ (X) = 0$ determines the interdependence between the components of the measurable parameters vector. The components of a vector stochastic process are, evidently, also constrained by this this dependence $Model\ (Z) = 0$. This interdependence, on the one hand, complicates the filtration problem and the structure of the synthesized filter, but, on the other hand, yields some additional information that improves the quality of filtration. Below we consider various static and dynamic mathematical models. The filtration is performed in such a way that model correlations survive in filtered values, i.e. $Model\ (L) = 0$.

## 3. Optimum filter structure

The following conventional signs will be used:

$i$ – vector component number $Z_i(j)$, $0 \leq i \leq I$,

$j$ – vector measurement instant $Z_i(j)$, $0 \leq j \leq J$,

$p$ – order of increment,

$\Delta^p$ – increment of the $p$-th order,

$M$ – mathematical expectation,

$0 \leq \beta_i \leq 1$ – weight coefficients,

$L_i(j)$ – optimum estimation of vector $Z_i(j)$ at instant $j$,

$E_i(j)$ – vector $Z_i(j)$ filtering error at instant $j$-момент, where

$$E_i(j) = Z_i(j) - L_i(j).\tag{1}$$

The first increment of the discrete stochastic process

$$\Delta^1 Z(j) = Z(j) - Z(j-1).$$

Increments of higher orders are found by the recurrent formula

$$\Delta^{p+1} Z(j) = \Delta^p Z(j) - \Delta^p Z(j-1).$$

The stochastic process $Z(j)$ with $p$-th stationary increments $\Delta^p$ is characterized by the fact that mathematical expectation

$$M\left(\Delta^{p+1} Z(j)\right) = 0.$$

During the process of filter synthesis there is a good reason to pursue a similar condition for the filter output signal, i.e.

$$M\left(\Delta^{p+1} L(j)\right) = 0.\tag{2}$$

Besides, it is necessary to try to meet the condition

$$\overline{M}\left(E^2(j)\right) = 0.\tag{3}$$

In this connection the following filtering quality characteristic has been chosen:

$$R = \overline{M} \sum_{i,j} [(1 - \beta_i)(\Delta^{p+1} L_i(j))^2 + \beta_i E_i^{\,2}(j)],\tag{4}$$

The filter structure is independent on $J$. Therefore, with the given $p$ and $\beta_i$ it can be pre-synthesized and used in real-time systems. An optimum filter in terms of minimum quality characteristic $R$, is

$$L(j) = B \cdot W(j),\tag{5}$$

where

$B$ – filter matrix,

$W(j)$ – detemined vector-function of measurements vector $Z(j)$ and values of vectors $Z(k)$ and $L(k)$ at past instants $k = j\text{-}1, j\text{-}2, ..., j\text{-}p$; this vector is of the following form:

$$W(j) = \begin{vmatrix} \Delta^p Z(j) \\ \Delta^{p-1} Z(j) \\ ... \\ \Delta^2 Z(j) \\ \Delta^1 Z(j) \\ Z(j) \\ L(j-1) \\ \Delta^1 L(j-1) \\ \Delta^2 L(j-1) \\ ... \\ \Delta^{p-1} L(j-1) \\ \Delta^p L(j-1) \end{vmatrix} \qquad (6)$$

The lentgth of this vector is
$$G = 2 * p + 2. \qquad (7)$$

## 4. Classification of RSSI filters

Table 1 contains a classification of stochastic processes and related filters, Table 2 contains major properties of these filters. They will be described in more detail in one of the subsequent sections of this work.

*Table 1.*

| process | | | |
|---|---|---|---|
| stationary | nonstationary | | |
| 1. | 2. | vector | |
| scalar | scalar | 3. simple (no model) | multiply correlated (see below) |

| Multiply Correlated Vector Process | | | | |
|---|---|---|---|---|
| Static Model | | | Dynamic Model | |
| Type 1 – with strong constraints | 6. Type 1 – with weak constraints | 7. Type 2 | 8. Type 1 | 9. Type 2 |
| 4. General case / 5. Synchronous processes | | | | |

*Table 2.*

| № | Process | $b$ | Filter | $S$ |
|---|---|---|---|---|
| 1 | Scalar stationary process, exponential smoothing filter, $p=0, I=1$ | 2 | (5), (7a) | $\beta$ |
| 2 | Scalar nonstationary process | G | (5) | $p, \beta$ |
| 3 | Vector process without a model | [I]*[G*I] | (5) | $p, I, \beta$ |
| 4 | Vector multiply correlated process with the static model of type 1 (8) | [I]*[G*I+U] | (5) | $A, C, p,$ $U, I, \beta$ |
| 5 | Synchronous processes | [I]*[G*I+U] | (5) | $p, I, \beta$ |
| 6 | Vector process with a static nonstrict model of type 1 (8), (12) | [I]*[G*I+U] | (5) | $A, D, p,$ $U, I, \beta$ |
| 7 | Vector process with the static model of type 2 (13) | [I]*[G*(I+U)] | (14) | $A, p, U,$ $I, \beta', \beta''$ |
| 8 | Vector process with a dynamic model of type 1 (16) | [I]*[G*(I+U)+U] | (17) | $A, p, U,$ $I, \beta', \beta''$ |
| 9 | Vector process with a dynamic model of type 2 (19) | [I]*[G*I+I] | (5) | $A, p,$ $I, \beta$ |

In Table 2:

    $b$ – dimension of matrix $B$,

    $I$ – dimension of measurements vector,

    $G$ – dimension of vector $W$ – s. (7),

    $U$ – number of object model equations,

    $S$ – data for synthesis

### 4.1. Scalar stationary processes

In this case we consider a single stochastic process at $p=0$, $I=1$. An optimum filter in terms of minimum quality characteristic (4), is of the form (5), where

$$W(j) = \begin{vmatrix} Z(j) \\ L(j-1) \end{vmatrix}, \; B = \begin{vmatrix} \beta & (1-\beta) \end{vmatrix}.$$

Thus,

$$L(j) = \beta \cdot Z(j) + (1-\beta)L(j-1) \tag{7a}$$

i.e. filter RSSI-0 is the same as the wide-spread exponential smoothing filter.

### 4.2. Scalar nonstationary processes

In this case we consider a single stochastic process at $p>0$, $I=1$. An optimum filter in terms of minimum quality characteristic (4), is of the form (5), where $G=2p+2$.

### 4.3. Uncorrelated vector processes

Let us consider a vector stochastic process with independent components at $I>1$. It is obvious that each component's filter is synthesized independently. If components have the same orders, then matrix B can be constructed for a vector process as a whole.

### 4.4. Vector process with the static model of type 1

In this case the static model takes the form

$$A \cdot Z(j) = C(j) \tag{8}$$

where

    $A$ – known matrix of dimension $U*I$,

    $C(j)$ – known vector that can be time variant

    $U$ – number of model equations and the dimension of vector $C$.

In this case filter (5) takes the form

$$L(j) = B \cdot W'(j), \tag{9}$$

where

$$W'(j) = \left| \begin{matrix} W(j) \\ C(j) \end{matrix} \right|, \tag{10}$$

i.e. the dimension of vector $W'(j)$ equals $I*G+U$, where $G=2p+2$. The dimension of matrix $B$ of filter (9) equals $[I]*[I*G+U]$.

For filtered vector values a condition similar to condition (8) is fulfilled:

$$A \cdot L(j) = C(j) \tag{11}$$

Let us remark here that matrix $B$ is independent on vector $C(j)$. Therefore,

|| vector $C(j)$ can be time variant.

## 4.6. Synchronous processes

Synchronous processes are characterized by the fact that between each pair of processes there remains the dependence $Z_a(j) - Z_b(j) = Y_k$, $k = func(a,b)$. A model of such processes is a special case of the static model of type 1, and matrix $A$ and vector $C$ are formed automatically at a given $I$. E.g., at $I=4$ we obtain: $U = 10$. Let us remark that in this case matrix $B$ is independent on vector $\{Y_k\}$. Therefore,

|| values $\{Y_k\}$ can be time variant.

| A | | | | | C |
|---|---|---|---|---|---|
| 1 | −1 | 0 | 0 | 0 | C[0]=Z[0]-Z[1]; |
| 1 | 0 | −1 | 0 | 0 | C[1]=Z[0]-Z[2]; |
| 1 | 0 | 0 | −1 | 0 | C[2]=Z[0]-Z[3]; |
| 1 | 0 | 0 | 0 | −1 | C[3]=Z[0]-Z[4]; |
| 0 | 1 | −1 | 0 | 0 | C[4]=Z[1]-Z[2]; |
| 0 | 1 | 0 | −1 | 0 | C[5]=Z[1]-Z[3]; |
| 0 | 1 | 0 | 0 | −1 | C[6]=Z[1]-Z[4]; |
| 0 | 0 | 1 | −1 | 0 | C[7]=Z[2]-Z[3]; |
| 0 | 0 | 1 | 0 | −1 | C[8]=Z[2]-Z[4]; |
| 0 | 0 | 0 | 1 | −1 | C[9]=Z[3]-Z[4]; |

($A=$ labels the matrix on the left)

In Figure 1 graphically shown is the result of the filtration of a synchronous 6-dimensional vector process with stationary increments of the second order, synchronous circular movement of three points (yellow curve). Here each stochastic process is changed coordinates of a point, on which noise is superimposed (blue curve). The filtered process (red curve) comes close to the ideal circle.

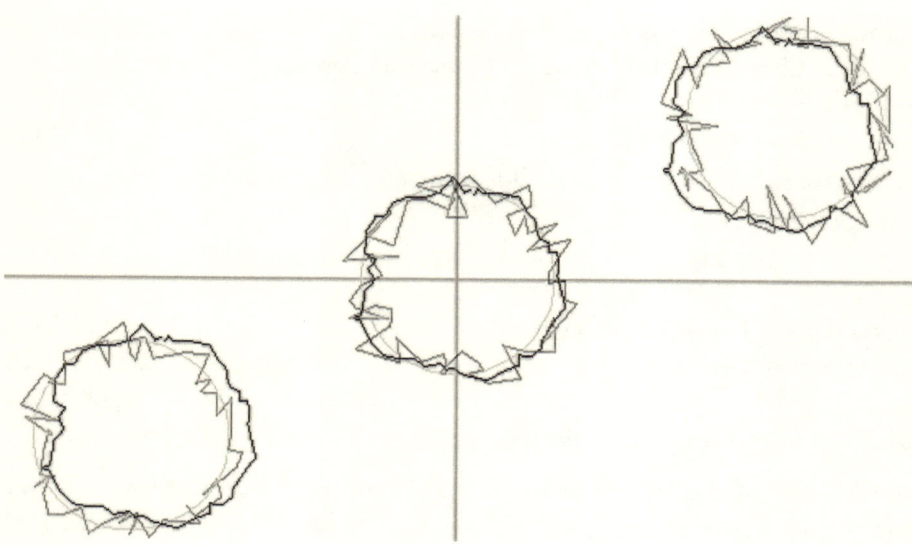

### 4.6. Static model of type 1 at weak constraints

For the static model of type 1 the quality coefficient can be supplemented and presented as

$$T = R + \overline{M}\sum_{j}[[A \cdot L(j) - C(j)]^T D \cdot [A \cdot L(j) - C(j)]], \qquad (12)$$

where $D$ is a diagonal matrix of the dimension $U * U$ of known weight coefficients $d_k$. At $d_k \gg 1$ the model equations are strongly fulfilled. At $d_k \approx 1$ the model equations can be violated to the same degree as filtering errors, i.e. conditions (2) and (3). Finally, at $d_k \approx 1$ and different $k$ the model equations can be fulfilled with different degrees of accuracy.

### 4.7. Vector process with the static model of type 2

The static model of type 2 takes the following form:

$$Z''(j) = A \cdot Z'(j), \qquad (13)$$

where

$Z''(j),\ Z'(j)$ – two different vectors (here primes identify two different RSSI-$p$'s and all values associated with them),

$U$ – number of  model equations and the dimension of
measurements vector $Z''(j)$,

$I$ – dimension of measurements vector $Z'(j)$,

$A$ – known model matrix of the dimension $U*I$.

An optimum filter in terms of quality characteristic (4) is

$$L'(j) = B \cdot W,$$
$$L''(j) = A \cdot L'(j),$$

(14)

where

$B$ – matrix of the dimension *[I]\*[G\*(I+U)]*,

$W$ – vector of the dimension $G*(I+U)$, where $G=2p+2$, and

$$W(j) = \begin{vmatrix} W'(j) \\ W''(j) \end{vmatrix}$$

(15)

## 4.8. Vector process with the dynamic model of type 1

The dynamic model of type 1 takes the form

$$Z''(j+1) = A \cdot Z'(j),$$

(16)

i.e., it connects nearest objects in terms of measurement time that belong to two different processes.

Here

$Z''(j)$, $Z'(j)$ – two different vectors (here primes identify two
different RSSI-$p$'s and all values associated with them),

$U$ – number of  model equations and the dimension of
measurements vector $Z''(j)$,

$I$ – dimension of measurements vector $Z'(j)$,

$A$ – known model matrix of the dimension $U*I$.

An optimum filter in terms of quality characteristic (4) is

$$L'(j) = B \cdot W,$$
$$L''(j+1) = A \cdot L'(j),$$

(17)

where

$B$ – matrix of the dimension *[I]\*[G\*(I+U)+U]*,

$W$ – vector of the dimension *[G\*(I+U)+U]*, where $G=2p+2$, and

$$W(j) = \begin{vmatrix} W'(j) \\ W''(j) \\ L''(j) \end{vmatrix}$$

(18)

### 4.9. Vector process with the dynamic model of type 2

The dynamic model of type 2 takes the form

$$Z(j+1) = A \cdot Z(j) \tag{19}$$

i.e., it connects nearest objects in terms of measurement time that belong to one process. Such an object can be one with a known movement model.

Here $A$ is a known square matrix of the dimension $I*I$. An optimum filter in terms of quality characteristic (4) takes the form (5), where

$B$ – matrix of the dimension $[I]*[G*I+I]$,

$W$ – vector of the dimension $[G*I+I]$, where $G=2p+2$, and

$$W(j) = \begin{vmatrix} W(j) \\ L''(j-1) \end{vmatrix} \tag{20}$$

# 5. Filtration procedure

On the completion of filter synthesis, i.e. calculation of matrix B, filtration of measurements is performed on-line by the following procedure:

✓ receipt of vectors Z(j), C(j);

✓ formation of new vector W(j) based on known vectors W(j-1), L(j-1), C(j), Z(j);

✓ calculation of a vector of smoothed L(j) based on known vectors W(j), B;

✓ repetition of the above calculation group for the next value of $j$, etc.

# 6. Programming technology

Programming of a RSSI-filter using the method proposed consists of the following stages:

• given process analysis and selection from *mathematical models of filters* of such a model that would be adequate to the given process;

• *filter synthesis* for the selected mathematical model; this stage is fulfilled by means of the proposed program;

• programming of a real-time filter; for this purpose a *function library* is proposed.

# 7. Conclusion

An indication can be made of a number of practically useful properties of the proposed method:

- optimality of a multivariate filter for nonstationary stochastic processes.
- feasibility at any order of stationary increments.
- applicability to real-time systems.
- applicability to cases where no information about signal and noise static properties is available.
- preservation for filtered values of a known dependence between RSSI components; this dependence can be static, i.e. time-independent, or dynamic, i.e. account for the association between component values at consequent instants of time;
- improvement in filtration quality  owing to the additional information provided by a known interdependence between RSSI's.

The proposed method can be used in real-time systems for controlling objects with determined mathematical models, e.g.:

- in power grid supervisory control systems, oil and gas pipelines (it is known that appropriate models are adequately described by a linear equation system);
- in industrial continuous process control (mathematical models for these processes can often be linarized);
- in recognition of objects whose shape changes affinely;
- in particular, in radar systems tracking multiple object groups (it can be shown that in such systems the process is described by equations resulting from the limited maneuvring of objects within a group; the criterion of filter optimality in this case  is a rephrased requirement for filtering error minimization at a group acceleration limited due to physical restrictions).

# References

1. Gelfand A.M., Khmelnik S.I. Filter synthesis for stationary processes with random increments, CAD for automated control system mathematical and informational support, Collection of scientific papers of the Central Scientific Research Institute for Complex Automation - CNIIKA, Moscow, Energoatomizdat, 1988.

2. Gelfand A.M., Khmelnik S.I. Filtration of correlated stochastic processes with stationary increments of a predetermined order, Automated control systems for chemical production, Collection of scientific papers of the Central Scientific Research Institute for Complex Automation - CNIIKA, M., Energoatomizdat, 1988.

3. Gelfand A.M., Khmelnik S.I. Mathematical modelling in estimation and filtration algorithm synthesis problems, Mathematical modelling of control objects, Collection of scientific papers of the Central Scientific Research Institute for Complex Automation - CNIIKA, Moscow, NPO CNIIKA, 1991.

4. Gelfand A.M., Khmelnik S.I. Development of digital filters for one class of nonstationary time series, Automated control systems, Mashinostroenie, Москва, 1994, No 2.

5. Tu Yu. Modern control theory. Moscow, Mashinostroenie, 1971, 472 p.

## Series: **Computer Engineering**

Solomon I. Khmelnik

# Positional codes of complex numbers and vectors

## Abstract

A little-known theory of positional coding of complex numbers and vectors is described; this theory may be used for the development of specialized processors. The theory is supplemented by numerous examples.

## Contents

## Introduction

To improve performance of complex number processing in computer systems several proposals have been made for a *single-component* representation. The essence of these proposals is to present complex numbers in such a positional number system, which allows for complex operations to be implemented on a single data element without a need for separate processing of real and imaginary components.

S. Khmelnik [2, 5, 9] was among the first to analyze and propose various positional codes for presentation and processing of complex numbers. He proposed and analyzed several positional coding systems, including those with $j\sqrt{2}$ radix and $(-1+j)$ radix. Other early researchers in this field include D. Knuth [1] who suggested to use an imaginary radix $j\sqrt{2}$ for positional complex codes and W. Penney [4] who afterward also suggested to use a complex radix $(-1+j)$. In his works [2,5,6,7,9,14,15] S. Khmelnik suggested techniques for coding and decoding complex numbers and apparatus for arithmetic and mathematic processing of complex numbers, and then in [3,8,14-21] – implementation of these operations in digital hardware. Last works by S. Khmelnik [14-21] are being implemented in specialized ALUs for complex number processing.

Later several authors [11-13] suggested techniques for design of complex number multipliers. They used redundant codes for complex number representation, in order to achieve more regularity in ASIC hardware. However these coding systems were never applied to implementation of other arithmetic operations, such as division.

This article only summarizes the information about various positional codes for complex numbers and vectors.

# 1. About the method of positional coding

In this section we shall discuss positional codes of many-dimensional vectors $Z$, based on their expansion in a series

$$Z = \sum_m r_m f(\rho, \ m), \tag{1}$$

where

$m$ - number of position

$\rho$ - radix of coding – a number or a vector,

$f(\rho, \ m)$ - base function of the number and the radix,

$r$ – the expansion's position, number or vector, assuming values from a bounded set $A_R = \{a_0, a_1, a_2, ..., a_j, ..., a_{R-1}\}$, containing $R$ various values $a_j$.

The positional code of a vector $Z$, respective to this expansion, has the form

$$K(Z) = ... \sigma_m ...,$$

where $\sigma_m$ - the digit that denotes the value $r_m$. Formula (1) includes operations of addition and multiplication. For existence of algorithms for operations with such expansions (or, which is the same, with positional codes) the operations of addition and multiplication should be associative and commutative, and also should obey the distributive law. Hence, for positional coding of some set of objects to be possible, this set should be a ring. The set of real numbers and the set of many-dimensional vectors, with determined operations of addition and multiplication by number, satisfies this requirement. For real numbers the positional systems are known. For the above indicated set of vectors a positional system with the real basis will be developed below.

The set of complex numbers is a ring, and for it positional number systems in real and complex radix will be also constructed.

For development of a positional number system for many-dimensional vectors in a vector radix, operation of vectors multiplication subject to the above named laws should be determined. In other words, an algebra in many-dimensional vector space should be determined. This will be accomplished in the following.

First we shall consider two ways of vectors coding; then we shall proceed to a more general and rigorous description of positional coding method.

## 2. Two ways of complex numbers codes synthesis

Positional codes of many-dimensional vectors may be constructed as a certain composition of codes of real numbers to negative radix. Here and further $j$ is imaginary unit.

Let $X_\alpha$ and $X_\beta$ - be real numbers, defined as binary expansions in the radix $\rho = -2$, that is

$$X_\alpha = \sum_{(m)} \alpha_m \rho^m , X_\beta = \sum_{(m)} \beta_m \rho^m .$$

The codes corresponding to these expansions are

$$K(X_\alpha) = \alpha_m, \qquad K(X_\beta) = \beta_m.$$

There are two ways of joining these two codes into a code of complex number. According to the **first** of them a pair of positions $\alpha_m$ and $\beta_m$ are notated by one digit $\sigma_m$. Thus a code

$$K(Z) = ...\sigma_m...$$

of complex number $Z = X_\alpha + jX_\beta$ in the radix $\rho = -2$ with positions that assume one of the four values. The resulting code will be as follows:

$$\sigma_m \in \{0, 1, j, 1+j\}.$$

Let us consider now a complex function of a real integer argument $m$:

$$\rho_2 = \begin{cases} (-2)^{m/2} & \text{if } m-\text{even} \\ j(-2)^{m-1/2} & \text{if } m-\text{odd} \end{cases} \tag{2}$$

The considered code of this complex number to the radix of (-2) with complex values of positions may be regarded as a code of complex number on the radix $(\rho_2)$ with binary positions. This code corresponds to the expansion of a complex number: $Z = \sum_m \left(\sigma_m \rho_2^m\right)$, where binary

positions are $\sigma_m = \begin{cases} \alpha_m & \text{if } m - \text{even} \\ j \cdot \beta_m & \text{if } m - \text{odd} \end{cases}$. For illustration we shall

show here the codes of some characteristic numbers in this system:

$K(2) = 10100$, $K(-2) = 100$, $K(-1) = 101$, $K(j) = 10$, $K(-j) = 1010$, $K(2j) = 101000$.

The **second** way consists in construction of a sequence of alternate positions $\alpha_m$ and $\beta_m$

$$...\beta_{m+1}\alpha_{m+1}\beta_m\alpha_m\beta_{m-1}\alpha_{m-1}...$$

Denoting $\alpha_m = \sigma_{2m}$, $\beta_m = \sigma_{2m+1}$, we shall rewrite this sequence in another form:

$$...\sigma_{k+3}\sigma_{k+2}\sigma_{k+1}\sigma_k\sigma_{k-1}\sigma_{k-2}$$

where $k=2m$. This sequence is the binary code

$$K(Z) = ...\sigma_m...$$

of a certain complex number $Z$. It is possible to show ( and it will be done below ), that the code obtained in such a way is a binary code in the radix $\rho = \pm j\sqrt{2}$, and the coded number is $Z = X_\alpha + \rho \cdot X_\beta$.

Thus, certain compositions of binary codes of real numbers in the radix $\rho = -2$ are the codes of complex numbers. When performing the algebraic addition of complex numbers such codes may be considered simply as a set of real numbers' codes and so we may perform the same operation independently with each pair of real numbers. Operations of multiplication for such codes and division for the codes of the second type are executable . The operations of multiplication and division consist, as usual, of cycles "shift - addition".

# 3. Method of coding for the points of many-dimensional space

A method of coding for points of many-dimensional Euclidian space should establish a certain correspondence between these points and the codes from a certain set. This correspondence, generally speaking, is not necessarily a one-to-one correspondence. But in order to make the decoding unique, each code should correspond to an unique point of the coded space. At the same time even a bounded part of the space contains an uncountable set of points. Hence, the set of corresponding codes is also uncountable, and among them there are bound to be codes with infinite number of positions (**infinite codes**). However in practice of calculations only **final codes** are used, and the set of final codes is bounded.

To preserve a correspondence between the codes and the points of space, it would appear reasonable to divide the bounded coded domain $G$ into a bounded set of sub-domains $\delta$ of a fixed size and configuration, so that each point of the domain $G$ will belong to one of sub-domains $\delta$. Then a one-to-one correspondence between set of final codes and set of the areas $\delta$ will be attainable.

Such way of coding the points of many-dimensional space is approximate. Indeed, a unique code $K_i$ corresponds to all points $Z_j \in \delta_i$. However when decoding the code $K_i$ we get an unique point $Z_i$. Denote the radius-vector of a point $Z$ by symbol $\overline{Z}$. The difference $\Delta Z_j = |\overline{Z}_j - \overline{Z}_i|$ defines the absolute error of point $Z_j$ coding.

For illustration let us consider fig. 1, where we see domain $Z_j$ of the two-dimensional space, divided into sub-domains $\delta$.

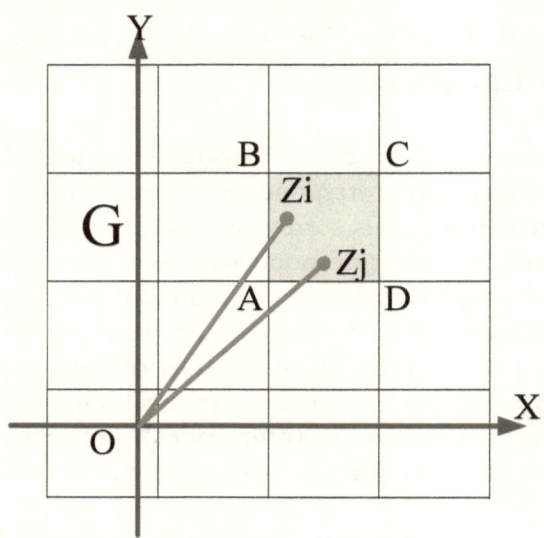

*Fig. 1. Coding of two-dimensional region*

On this figure one of the sub-domains $\delta_i$ = ABCD is marked out, where its lower (AD) and right (CD) borders also belong to $\delta_i$. In $\delta_i$ we have taken a basic point $Z_i$ and another point $Z_j \in \delta_i$. The length of segment $\Delta Z_j$ characterizes the absolute error of the point $Z_j$ coding.

So, the stated principle of coding for points of many-dimensional space may be formulated as following:

- bounded domain $G$ of the coded space is divided into a bounded set of equal sub-domains $\delta_i$ ($i=1, 2,..., N$), where

$$G = \bigcup \delta_i \text{ and } \delta_i \bigcap \delta_j = \varnothing \text{ at } i \neq j ;$$

- a set of final codes $K_i$ ($i=1, 2,..., N$) is defined;
- between the sub-domains and the codes a one-to-one correspondence is established.

If these conditions are observed, it means that the **system of coding** of the domain $G$ of many-dimensional space satisfies the **principle of coding** and the domain $G$ is coded with **discreteness** $\delta$. The following two lemmas are obvious.

**Lemma 1**. The system of coding of domain $G$ satisfies the principle of coding if $V=NU$, and vice versa, where

$U$ – is the $\delta$ -domain's volume,

$V$ – is the $G$ –domain's volume,

$N$ - the power of final codes set.

**Lemma 2**. The system of coding for $G$ - domain, satisfying the principle of coding, is called **complete** (that is, any point corresponds to a final code), **nonredundant** (that is, each point corresponds to one final code) and **approximate** (that is, a subset of points - vectors, for which module of difference does not exceed a certain value, corresponds to one final code).

Consider a set of $n$-digit codes of the form

$$K = \alpha_{n-1}...\alpha_k...\alpha_1\alpha_0 , \qquad (3)$$

where $\alpha_k$ is a digit that assumes one of the values $R_k$, $R_k > 1$ and is integer.

**Lemma 3**. If the system of coding satisfies the principle of coding, then with increase of final codes digit capacity and on retention of coding discreteness, the coding volume of the coded domain is increased in the same way as the power of final codes set, and vice versa.

*Proof*. The power of final codes set is

$$N_n = \prod_{k=1}^{n} R_k . \qquad (4)$$

Let this set of codes satisfy the principle of coding and encode the domain $G_n$ with discreteness $\delta$ . According to lemma 1, the number of sub-domains $\delta$ , contained in $G_n$, is also equal to $N_n$, and the $G_n$ domain has volume

$$V_n = N_n U . \qquad (5)$$

Now we shall increase the codes digit capacity by 1, that is, we add a digit $\alpha_n$, which assumes one of the values $R_n$. It is obvious that

$$N_{n+1} = R_n N_n . \qquad (6)$$

Let the new set of codes also satisfy the principle of coding and encode the domain $G_{n+1}$ with the same discreteness $\delta$ . The number of sub-domains $\delta$ , contained in domain $G_{n+1}$, is equal to $N_{n+1}$, that is, the domain $G_{n+1}$ has volume

$$V_{n+1} = N_{n+1} U . \qquad (7)$$

Combining three last formulas, we find, that

$$V_{n+1} = R_{n+1} V_n , \qquad (8)$$

that is, the direct part of the lemma is proven.

By the conditions of the inverse part of lemma, the formulas (5), (6), (8) are valid. From them follows (7), whence according to the lemma 1 we come up with the proof of the inverse part of the lemma.

We shall consider now a positional system of coding. In this system each positional code

$$K(Z) = \alpha_n ... \alpha_k ... \alpha_m$$

corresponds to a point $Z$ of a coded many-dimensional space, which has an expansion of the following form:

$$Z = \sum_{k=m}^{n} \alpha_k \rho^k, \tag{9}$$

where

$\rho$ - is the radix of coding,

$k$ – the position number,

$\alpha_k$ - the $k$-th position of a code (digit or quantitative equivalent, corresponding to it in the expansion), which assumes one of the values $R_k$.

Notice that $\rho$ and $\alpha_k$ are also points of the coded many-dimensional space. The positional code is called **infinite** if $m = -\infty$, and **final** if $m$ is limited. Number $n$ called the **length** of positional code. If $R_k = R$, the expansion and the code are called $R$-expansion and $R$-code. So, we shall consider the value $a$, assuming values from the set

$$A_R = \{a_0, a_1, a_2, ..., a_j, ..., a_{R-1}\}, \tag{10}$$

containing $R$ different values $a_j$. In practice of positional coding it is essential that $R$ is limited and is not larger than a few units.

We shall denote the positional code of a point $Z$ on the radix $\rho$ and write it also as follows

$$< Z >_\rho = \alpha_n ... \alpha_k ... \alpha_1 \alpha_0, \alpha_{-1} \alpha_{-2} ... \alpha_m, \tag{11}$$

placing the point between the digits zero and (-1) ( index - base will not be indicated, if the base'd value is clear from the context). A vector (point) $Z$, in whose code $m \geq 0$, will be called $\rho$ - **whole**. Accordingly the vectors $Z$ that are $\rho$ - **fractional** ( **proper** and **improper** ) are defined. In particular,

$$< \rho >_\rho = 10. \tag{12}$$

The aggregate of $< \rho, \ A_R >$ of the radix of coding $\rho$ and the set $A_R$ shall be called a **system of positional coding**. We shall say that a point of many-dimensional Euclidian space is **representable** in the given system of positional coding, if there is a corresponding expansion of the form (9) and a positional code of the form (11), where the digits assume values from the set (10).

We have to construct such positional systems of coding, in which any point of the given space will be representable, subject to conditions of completeness, nonredundance and approximateness, according to lemma 2.

The purpose of positional systems construction is to simplify the execution of arithmetic operations with points ( vectors ) of the many-dimensional space. On the other hand, the existence of positional codes, based on expansion (9), is possible only for such space, where operations of summation of vectors and multiplication of a vector by the radix $\rho$ ( which can also be a vector ) are determined.

In one- and two-dimensional spaces multiplication by the radix $\rho$ (multiplication by real or complex number) corresponds to an increase of a module vector-multiplicand by the factor $|\rho|$, that is,

$$\text{if } Z_2 = Z_1\rho, \text{ then } |Z_2| = |Z_1||\rho|. \tag{13}$$

It should once again noted, that multiplication $Z_1\rho$ is equivalent to a shift of the code $<Z_1>$ by one position to the left *in any space*. We shall require the condition (13) to be valid also *for any coded space* and we shall prove a certain condition of a positional system's existence, using these two facts.

**Theorem 1**. For any point of an $h$-dimensional Euclidian space in which condition (13) is satisfied, the necessary and sufficient condition of its representability in the given system of positional coding is the condition:

$$|\rho|^h = R. \tag{14}$$

*Proof.* Each code $<Z_2>_\rho$ of length $(n + h)$ with $m = -\infty$ may be received by shift by $h$ positions to the left of a certain code $<Z_1>_\rho$ of length $n$. But according to (12) such shift is equivalent to multiplication by the radix, that is, $Z_2 = Z_1\rho^h$. Thus from formula (13) follows, that $|Z_2| = |Z_1||\rho|^h$. Hence, the linear sizes of the coded domain are increased by $|\rho|^h$ (besides the coded area, generally speaking, is turned relative to its previous position ). Thus, the volumes of the areas $G_n$ and $G_{n+h}$ are related by the following equation:

$$V_{n+h} = |\rho|^h V_n. \tag{15}$$

Obviously, the restriction $m$ does not change volume of the coded domain. There only appears discreteness of coding $\delta = G_{m-1}$. Taking into account (14), from (15) we obtain

$$V_{n+h} = RV_n. \tag{16}$$

Comparing (16) and (8), from lemma 3 we find that the system of positional coding at $m < -\infty$ satisfies the principle of coding, that is, owing to lemma 2, it is total, nonredundant and approximate. The theorem is proven.

## 4. Arithmetic systems of coding

Among positional systems of coding, such systems for which simple algorithms of addition and multiplication are applicable – are subjects of particular interest. Such indeed are the systems which we shall consider below, but first we must define them more strictly.

**Definition 1.** System $< \rho \quad , \quad A_R >$ of positional coding is called **arithmetic**, if following conditions are fulfilled
- number (-1) is $\rho$- whole,
- the sum and the product of any pairs of vectors, belonging to to set $A_R$, are $\rho$- whole.

Note, that the condition (13) may be valid also for a non-arithmetic system.

**Lemma 4.** If in an arithmetic positional system the vectors $Z_1$ and $Z_2$ are representable, then the vectors $-Z_1$, $-Z_2$, $Z_1 + Z_2$, $Z_1 Z_2$ are also representable in this system.

The lemma's validity follows from the fact that, as it will be shown below, for arithmetic positional systems there exist algorithms of arithmetic operations.

**Definition 2.** Positional system $< \rho, \quad A_R >$ is called **normal**, if $A_R = B_R$, where $A_R = B_R$, где $B_R = \{0,1,2,...,R-1\}$.

**Lemma 5.** A normal number system, in which

$$R = \sum_{k=1}^{n} \alpha_k \rho^k, \tag{17}$$

$$-R = \sum_{k=1}^{w} \beta_k \rho^k, \qquad (18)$$

that is, the codes of numbers $R$ and $-R$ are $\rho$-whole and have zero value of the zero digit, is an arithmetic system.

*Proof.* For any number from set $B_R$   $0 \le a_j \le (R-1)$. Hence, for numbers from this set the equations $-a_j = a_k - R$ and $a_j + a_k = a_m + R$, if $a_j + a_k \ge R$ are valid. Taking into account the conditions of the lemma, we conclude, that the numbers $(-a_j)$ and $(a_j + a_k)$ are $\rho$-whole. Obviously, the product $a_j a_k$ can be presented as a sum of numbers from the set $B_R$. By induction, owing to the existence of addition algorithm, we conclude, that such sum is also $\rho$-whole. Thus, the conditions of definition 1 are fulfilled. Hence, the considered system is an arithmetic one.

**Lemma 6.** A normal number system, in which the number R has an expansion of the form (17), and

$$R = \sum_{k=1}^{m} \alpha_k, \qquad (19)$$

is  an arithmetic system.

*Proof.* As follows from (17) and (19), in the lemma systems, in which

$$R = \sum_{k=1}^{n} \alpha_k \rho^k = \sum_{k=1}^{n} \alpha_k.$$

are considered.

Let us consider the following algorithm:

| | | | | | | | |
|---|---|---|---|---|---|---|---|
| $\alpha_3$ | $\alpha_2$ | $\alpha_1$ | 0 | | | | carries |
| | $\alpha_3$ | $\alpha_2$ | $\alpha_1$ | 0 | | | carries |
| | | $\alpha_3$ | $\alpha_2$ | $\alpha_1$ | 0 | | carries |
| | | | $\alpha_3$ | $\alpha_2$ | $\alpha_1$ | $0 =< R >_\rho$ | addend 1 |
| | | | | $\beta_2$ | $\beta_1$ | $0 =< X >_\rho$ | addend 2 |
| **0** | **0** | **0** | **0** | **0** | **0** | **0** | sum |

Here the code of number $R$ is summed the with code of some number $X$, whose digits are formed so that

$$\alpha_1 + \beta_1 = R \text{ и } \alpha_1 + \alpha_2 + \beta_2 = R.$$

Thus, and owing to (19) the addition of digits for each column will give us number $R$, which forms the carry and zero digit of the sum. As a result the infinite carries and the zero sum will be formed. Hence, $X=-R$. Obviously, such algorithm of number $-R$ code formation is executable for any $R$ corresponding to the expansion (17) or, which is the same, for any code of number $R$ of a form

$$< R >_\rho = \alpha_m...\alpha_2\alpha_1 0.$$

The result of this algorithm is a code of number $-R$:

$$< -R >_\rho = \beta_w...\beta_2\beta_1 0.$$

This code corresponds to expansion (18). Thereby the lemma is proved.

Note that the expansions (17) and (18) can be considered as a system of two power equations with unknown $\rho$. Solving it, we may, generally speaking, define a certain system of coding. However this method does not always lead to good results, because the given system either is not solvable analytically, or is not compatible, or gives a solution not satisfying the condition of Theorem 1, or gives a real number as the solution.

Lemmas 4, 5, 6 will be used further in the search for normal positional systems of coding.

## 5. Codes of real numbers

For real numbers the dimension of the coded space is $h=1$. Hence, for positional codes of real numbers it is necessary to observe a condition

$$|\rho| = R.$$

Positional codes of real numbers in which $\rho = R$ and the positions assume values from set $B_R$ are widely known and widespread. Among these are usual decimal $(R = 10)$ and binary $(R = 2)$ codes. However such codes can't portray negative real numbers, so we have to use some workarounds, in particular, inverse or additional codes, which may cause certain inconvenience.

At the same time there are two ways of developing positional codes fit to portray real - positive and negative numbers. First of them consists in giving positive and negative values from the set $D_R = \{-r_1, -r_1 + 1,...,-1,0,1,...,r_2 - 1, r_2\}$ to the positions $R = r_1 + r_2 + 1, r_1 \neq 0, r_2 \neq 0$, leaving the radix equal to $R$ (at $r_1 = 0$ the set $D_R$ turns into set $B_R$). The other way is based on the use of negative radix $\rho = -R$. Thus the digits may assume values either from set $B_R$, or from set $D_R$. So the known results related to positional coding of real numbers, may be formulated as follows.

**Theorem 2**. Any real positive number may be represented in the systems

$$< R, B_R >, \quad < R, D_R >, \quad < -R, B_R >, \quad < -R, D_R > .$$

So, there are four systems of real numbers coding:

the system $< R, B_R >$, for example < 5, { 0, 1, 2, 3, 4 } >;

the system $< R, D_R >$, for example < 5, { -2, -1, 0, 1, 2 } >;

the system $< -R, B_R >$, for example < -5, { 0, 1, 2, 3, 4 } >;

the system $< -R, D_R >$, for example <-5, {-2, -1, 0, 1, 2 }>.

We shall give some examples of penta-codes of numbers in the above-named systems, denoting the values -1 and -2 as $\overline{1}$ and $\overline{2}$:

1. K(16)= +31,K(-13)= -23,
2. K(16)= $1\overline{2}1$,K(-13)= $\overline{1}22$,
3. K(16)= 121,K(-13)= 32,
4. K(16)= 121,K(-13)= $\overline{1}\overline{2}2$.

Here we must draw attention to the fact that in the first of these systems the codes have " + " and "-", which are absent in all the other systems, for in them the number's sign, as well as the module, are defined by the values of code's digits.

It is important to note, that among indicated systems there are only two systems of binary coding, namely the system with numbers { 0, 1 } and radixes "2" and "-2".

## 6. Codes of complex numbers

We shall begin with proving some existence theorems of normal arithmetic systems of coding with complex radix, denoting imaginary unit by $j$.

**Theorem 3**. Any complex number is representable in a normal system of coding in a complex radix $\rho$, and this system is arithmetic, if

$$| \rho | = \sqrt{R} \tag{20}$$

and conditions (17), (19) are fulfilled.

*Proof.* For complex numbers the dimension of the coded space is $h=2$, and at any $\rho$ the condition (13) is fulfilled. From here and from (20) follows, that the conditions of theorem 1 are fulfilled. Hence, any complex number is representable in the given system of coding. Further, the conditions (17) and (19) are the conditions of lemma 6. Hence, the given system is arithmetic.

Theorem 3 enables to reduce the proof of theorems about representability of any complex number in a normal system of coding and arithmeticality of this system to the proof that condition (19) is satisfied and $\rho$ is a complex root of the equation (17). This very method of proof we shall use in future.

**Theorem 4**. Any complex number is representable in a normal system of coding at complex radix

$$< \rho = \sqrt{2}e^{\pm j\pi/2}; \ B_2 > \text{ or } < \rho = -1 \pm j; \ \{0, \ 1\} >$$

and this system is arithmetic.

*Proof.* Let us assume, that $< 2 >_\rho = 1100$. This condition it is equivalent to an equation $\rho^3 + \rho^2 = 2$. Its decision coincides with the condition of the given theorem. Hence, condition (17) is satisfied. Obviously the condition (19) is also satisfied, because $R=2$. In view of theorem 3 the given the theorem is proven. For illustration let us write several characteristical numbers in a system at the radix $\rho = (j-1)$, denoting by $\overline{\rho}$ - number adjoined to number $\rho$:

$K(2) = 1100$, $K(-2) = 11100$, $K(-1) = 11101$, $K(j) = 11$,
$K(-j) = 111$, $K(\overline{\rho}) = 110$.

**Theorem 5**. Any complex number is representable in a normal system of coding at a complex radix $\rho$ and this system is arithmetic, if

$$\rho = \sqrt{R}e^{j\varphi}, \quad \varphi = \pm \ arcCos \ (-\beta /2\sqrt{R}), \quad \beta < (R, \ 2\sqrt{R})_{min}$$

and $\beta$ - a whole positive number.

*Proof.* Let us assume that $< R >_\rho = 1\alpha_2\alpha_1 0$, where $1 + \alpha_2 + \alpha_1 = R$, $\alpha_2 = \beta - 1$. This condition is equivalent to an equation $\rho^3 + (\beta-1)\rho^2 + (R - \beta) \ \rho = R$. Its solution gives the value that was mentioned in the theorem's conditions. In view of theorem 3 the given theorem is proven.

For illustration let us record codes of some characteristic numbers in this system, denoting by $\overline{\rho}$ - number adjoined to number $\rho$:

$$K(R) = 1 \ (\beta-1) \ (R-\beta) \ 0,$$
$$K(-R) = 1 \ \beta \ 0,$$

$$K(-1) = 1 \ \beta \ (R-1),$$
$$K(\overline{\rho}) = 1 \ (\beta - 1) \ (R - \beta),$$
$$K(-\overline{\rho}) = 1 \ \beta,$$
$$K(-\rho) = 1 \ \beta \ (R-1) \ 0,$$
$$K(\rho - \overline{\rho}) = 2 \ \beta,$$
$$K(\rho + \overline{\rho}) = 1 \ \beta \ (R - \beta).$$

As $\beta$ can assume several values with constant $R$, there are several types of positional codes in systems of the considered type. As an example, possible codes of number $R$ with various $R$ and $\beta$ are presented in Table 1.

*Table 1. Codes of number R.*

| $R \setminus \beta$ | 1 | 2 | 3 | 4 | 5 |
|---|---|---|---|---|---|
| 2 | 1010 | | | | |
| 3 | 1020 | 1110 | | | |
| 4 | 1030 | 1120 | 1210 | | |
| 5 | 1040 | 1130 | 1220 | 1310 | |
| 6 | 1050 | 1140 | 1230 | 1320 | |
| 7 | 1060 | 1150 | 1240 | 1330 | 1420 |
| 8 | 1070 | 1160 | 1250 | 1340 | 1430 |
| 9 | 1080 | 1170 | 1260 | 1350 | 1440 |

By way of illustration we shall present the codes of several characteristic numbers in a system in the radix $\rho = \dfrac{1}{2}\left(-1 + j\sqrt{7}\right)$, denoting by $\overline{\rho}$ - number adjoined to number $\rho$:

$K(2) = 1010$, $K(-2) = 110$, $K(-1) = 111$, $K(\overline{\rho}) = 101$, $K(-\rho) = 1110$, $K(-\overline{\rho}) = 11$, $K\left(j\sqrt{7}\right) = 10101$, $K\left(-j\sqrt{7}\right) = 1110011$.

From the systems of theorem 5 it is possible to single out the groups with a fixed value of the base's argument, for example

$\varphi = \pm 2\pi / 3$, if $\beta = \sqrt{R}$, that is, at R=4, 9, 16, 25, ...;

$\varphi = \pm 3\pi / 4$, if $\beta = \sqrt{2R}$, that is, at R=8, 18, 32, 50,...;

$\varphi = \pm 5\pi / 6$, if $\beta = \sqrt{3R}$, that is, at R=12, 27, 48, 75,..;

Let us consider now a positional system of a more general form.

**Theorem 6**. Any complex number is representable in a system of coding $< \rho = 2e^{j\pi/3}$, $A_4 >$, $A_4 = \{0, 1, ,e^{2j\pi/3}, e^{-2j\pi/3}\}$ and this system is arithmetic.

*Proof.* Notice, that $(-2)^k = l_k \rho^k$, where

$$l_k = \left\{ 1, \ e^{2j\pi/3}, \ e^{-2j\pi/3} \right\}$$

accordingly at $k = (3m, 3m+1, 3m+2)$, where $m$ is integer. Obviously, $l_k \in A_4$. Hence, any power of the number "-2" may be represented in the indicated system of coding by one digit. In view of theorem 2 any real number $X$ is representable as an expansion in the radix "-2". But each digit of such expansion, being a power of number "-2" or 0, can be replaced by the term of expansion in the indicated system of coding, that is, any real number is representable in this system of coding.

*Table 2. One-digit multiplication*

| * | 0 | 1 | c | d |
|---|---|---|---|---|
| **0** | 0 | 0 | 0 | 0 |
| **1** | 0 | 1 | c | d |
| **c** | 0 | c | d | 1 |
| **d** | 0 | d | 1 | c |

Any complex number $Z$ can be presented as $Z = u_1 + u_2 e^{2j\pi/3} + u_3 e^{-2j\pi/3}$, where $u_1$, $u_2$, $u_3$ - certain real numbers. In this sum all the components are representable in the indicated number system, because the cofactors of real numbers $u_1$, $u_2$, $u_3$ belong to the set $A_4$. If this system is arithmetic, then the considered sum is representable in it, and so this is true for any complex number. It remains to show that the considered system is arithmetic. For that we shall arrange tables of pairwise multiplication, summation and the table of inverting ( multiplication by "-1" ) for figures from set $A_4$ - see tables 2, 3, 4. For convenience these figures are denoted by symbols 0, 1, c, d. As may be seen from these tables, in the considered system of coding all conditions of definition 1 are fulfilled. Hence this system is arithmetic, as it was required to show.

Table 3. One-digit summation

| + | 0 | 1 | c | d |
|---|---|---|---|---|
| **0** | 0 | 1 | c | d |
| **1** | 1 | dc0 | 1d | dc |
| **c** | c | 1d | d10 | c1 |
| **d** | d | dc | c1 | c10 |

Table 4. Inverting a digit

| x | 0 | 1 | c | d |
|---|---|---|---|---|
| -x | 0 | c1 | dc | 1d |

We must note, that in this system the codes of complex numbers of the forme $e^{jk60°}$ with integer $k$ have a very simple form - see table 4a. Besides, the table 4b presents the codes of numbers $2^k$ and $(-2)^k$ with integer $k$.

Table 4a. Codes of numbers $e^{jk60°}$ .

| φ | 0 | 60 | 120 | 180 | 240 | 300 |
|---|---|---|---|---|---|---|
| код | 00 | 1d | 0c | c1 | 0d | dc |

Now we shall only state more accurately the results obtained in section 2.

**Theorem 7**. Any complex number $Z$ is representable in a positional number system $< \rho = -R, \ A_{R^2} >$, where the set $A_{R^h}$ consists of complex numbers $r_m = \alpha_m^1 + j\alpha_m^2$, and the numbers $\alpha_m \in B_R$.

In particular, there exists a system $<-2, \{0,1,j,1+j\}>$.

*Table 4b. Codes of numbers $2^k$ and $(-2)^k$.*

| k | $(-2)^k$ | $2^k$ |
|---|---|---|
| -4 | 0.000d | 0.000d |
| -3 | 0.001 | 0.0c1 |
| -2 | 0.0c | 0.0c |
| -1 | 0.d | 1.d |
| 0 | 1 | 1 |
| 1 | c0 | dc0 |
| 2 | d00 | d00 |
| 3 | 1000 | c1000 |
| 4 | c000 | c000 |

**Theorem 8**. Any complex number $Z$ is representable in a normal positional system $< \pm j\sqrt{R}, B_R >$.

For example, there exists a system $< \pm j\sqrt{2}, \{0,1\} >$. For illustration we shall write the codes of several characteristic numbers in the system $\rho = j\sqrt{2}$:

$$K(2)=10100, \ K(-2)=100, \ K(-1)=101, \ K(j\sqrt{2})=10,$$
$$K(-j\sqrt{2})=1010.$$

*Table 5. Binary systems of coding.*

| Preferred number systems | $\rho$ | <2> | <-2> | <-1> | Theorem | Fig. |
|---|---|---|---|---|---|---|
| System 1 | $\rho_2$ | 10100 | 100 | 101 | Formula (2) | 1 |
| System 2 | $j\sqrt{2}$ | 10100 | 100 | 101 | Theorem 8 | 2 |
| System 3 | $-1+j$ | 1100 | 11100 | 11101 | Theorem 4 | 3 |
| System 4 | $\frac{1}{2}(-1+j\sqrt{7})$ | 1010 | 110 | 111 | Theorem 5 | 4 |
| | $-2$ | 110 | 10 | 11 | Theorem 2 | |
| | $2$ | 10 | | | Theorem 2 | |

Obviously, for the systems from theorems 7 and 8 the condition (14) is satisfied. The proof of these theorems is based on the reasonings of section 2.

For illustration and comparison let us present the binary codes of numbers in all the considered systems of coding, including systems of coding in real (positive and negative) and complex radix - see table 5.

Further we shall dwell in more detail on the four binary systems of complex numbers - see the column «Preferred number systems» in table 5. The following figures present the first 4 values of base function for the preferred number systems.

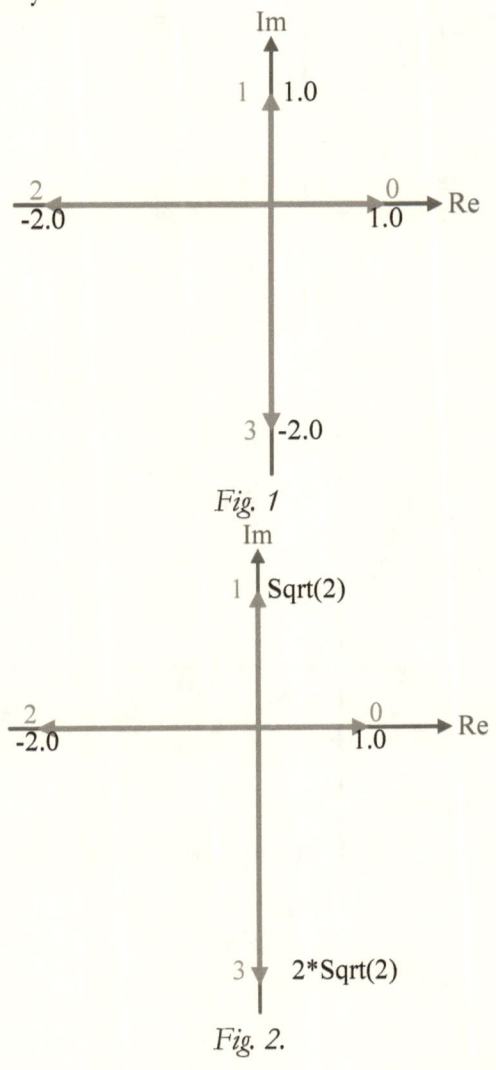

Fig. 1

Fig. 2.

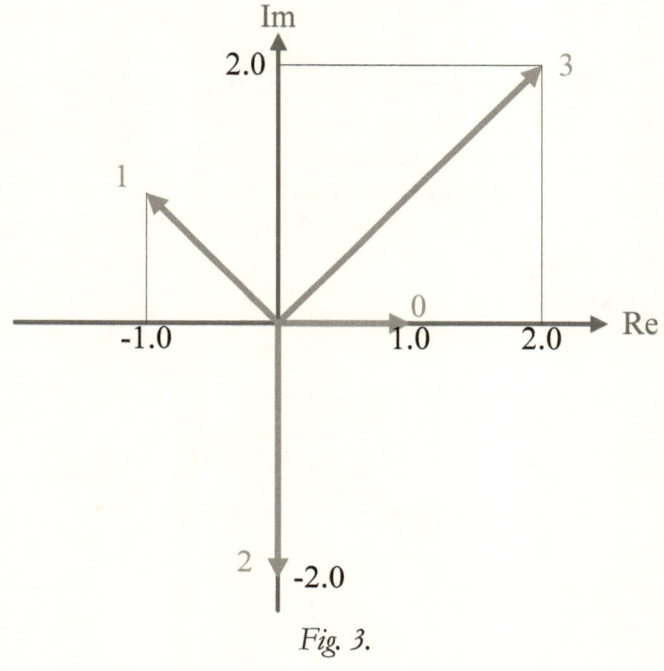

Fig. 3.

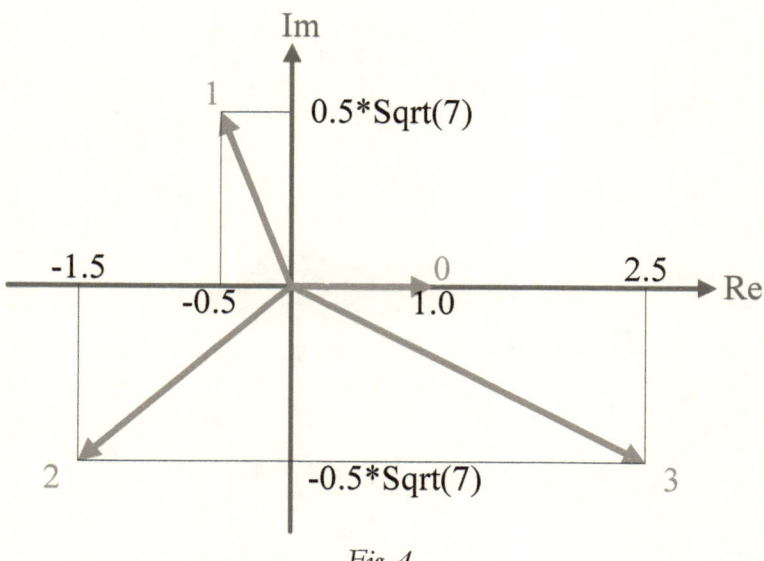

Fig. 4.

We shall also present a table of 6 four-valued codes of numbers ' 4 ' and '-4 ' in all systems of coding considered above (in this table '-1' it is designated by sign 'h').

Table 6. Four-valued coding system.

| $\rho$ | $A_4$ | <4> | <-4> | Theorem |
|---|---|---|---|---|
| 4 | {0,1,2,3} | 10 | | 2 |
| 4 | {-1,0,1,2} | 10 | h0 | 2 |
| 4 | {-2,-1,0,1} | 10 | h0 | 2 |
| -4 | {0,1,2,3} | 130 | 10 | 2 |
| -4 | {-1,0,1,2} | h0 | 10 | 2 |
| -4 | {-2,-1,0,1} | h0 | 10 | 2 |
| $2e^{2j\pi/3}$ | {0,1,2,3} | 1120 | 120 | 5 |
| $2e^{j\pi/3}$ | {0,1,c,d} | d00 | 1d00 | 6 |
| -2 | {0,1,j,1+j} | 100 | 1100 | 7 |
| $\rho_4$ | {0,1,2,3} | 10300 | 100 | 8 |
| $\pm 2j$ | {0,1,2,3} | 10300 | 100 | 8 |

# 7. Codes of many-dimensional vectors

### 7.1. Binary codes of vectors - method 1.

The stated method of complex numbers codes construction may be generalized and used for many-dimensional vectors coding. To do so we shall consider a set of real numbers $\{X_i\}$, each of which is described by a binary expansion in the radix $\rho = -2$, that is

$$X_i = \sum_{(m)} \alpha_m^i \rho^m \ (i=1, 2,..., n).$$

To each such expansion there corresponds a code

$$K(X_i) = ...\alpha_m^i ...$$

We shall consider now $n$-dimensional vector

$$Z = E_1 X_1 + E_2 X_2 + ... + E_n X_n, \tag{21}$$

where $\{E_i\}$ is the base of $n$-dimensional vector space. The set of codes $\{K(X_i)\}$ thus may be interpreted as an uniform code of the vector $Z$ in the radix "-2". Each m-th digit of this code is represented by the set

$\{\alpha\,_m^{\,i}\}$ of binary digits. Denoting these sets by digits $\sigma_m$, we get the code of the vector

$$K(Z) = ...\sigma_m...,$$

corresponding to the expansion (1), where the vector

$$r_m = E_1\alpha_m^1 + E_2\alpha_m^2 + ... + E_i\alpha_m^i + ... + E_n\alpha_m^n \qquad (22)$$

is represented by digit $\sigma_m$.

In particular, at $n = 2$ we shall get the codes of complex numbers in radix "-2," which have been considered higher. At $n=3$ we shall get the codes of three-dimensional vectors, where the digits take one of eight values:

$$r_m \in \{\, 0,\, i,\, j,\, k,\, i+j,\, i+k,\, j+k,\, i+j+k \,\}, \qquad (23)$$

where **i, j, k** are orts of Cartesian coordinate system. Similarly, for coding three-dimensional vectors, a function from real whole vector argument $m$ may be introduced:

$$\vartheta_2 = \begin{cases} i(-2)^m & \text{if } m = 3k \\ j(-2)^{m-1} & \text{if } m = 3k+1 \\ k(-2)^{m-2} & \text{if } m = 3k+2 \end{cases}, \qquad (24)$$

The considered code of a three-dimensional vector in the radix $\vartheta_2$ ) with vector digit values (23) may be considered a code of a three-dimensional vector in the radix $\vartheta_2$ with binary digits. To this code there corresponds the vector's expansion $Z = \sum\limits_{m}\left(\alpha_m\vartheta_2\right).$

Similarly, for $n$-dimensional vectors coding a vector-function of real integer argument $m$ may be introduced:

$$\vartheta_2^n = \begin{cases} i(-2)^m & \text{if } m = nk \\ j(-2)^{m-1} & \text{if } m = nk+1 \\ ... \\ k(-2)^{m-n+1} & \text{if } m = nk+n-1 \end{cases}, \qquad (25)$$

Obviously, $\rho_2 = \vartheta_2^2, \ \vartheta_2 = \vartheta_2^3.$

### 7.2. Binary codes of vectors - method 2.

We shall now build, as we have done earlier for complex numbers, a sequence of alternated binary digits $\alpha_m^i$:

$$\cdots\alpha_{m+1}^2\alpha_{m+1}^1\alpha_m^n\alpha_m^{n-1}\cdots\alpha_m^2\alpha_m^1\alpha_{m-1}^n\alpha_{m-1}^{n-1}\cdots$$

In other notations this sequence is a binary code

$$K(Z) = \ldots\alpha_k\ldots\ldots$$

of a certain vector $Z$. The radix of coding is also a vector

$$\rho = E_2\sqrt[n]{2}, \tag{26}$$

where $E_2$ - the second ort of the base $\{E_i\}$ of an $n$-dimensional vector space. The coded vector $Z$ is defined in this case by the formula

$$Z = X_1 + \rho X_2 + \ldots + \rho^{i-1}X_i + \ldots + \rho^{n-1}X_n. \tag{27}$$

### 7.3. Many-dimensional codes of vectors - method 2.

The positional codes of vectors (including complex numbers and multidimensional vectors) are built precisely similarly, based on incorporation of positional codes of numbers - projections of vectors at the radix $\rho = -R$, where $R$ is integer. In this case, for example, instead of the function $\rho_2$ the function

$$\rho_R = \begin{cases} (-R)^{m/2} & \text{if } m-\text{even} \\ j(-R)^{m-1/2} & \text{if } m-\text{odd} \end{cases}, \tag{28}$$

should be considered as the radix of complex numbers coding, and instead of function $\vartheta_2$ the function

$$\vartheta_R = \begin{cases} i(-R)^{m/3} & \text{if } m = 3k \\ j(-R)^{m-1/3} & \text{if } m = 3k+1 \\ k(-R)^{m-2/3} & \text{if } m = 3k+2 \end{cases}, \tag{29}$$

should be considered as the radix of complex numbers coding, and, generally speaking, instead of function $\vartheta_2^n$ the function $n$-dimensional vectors coding we should consider the following function

$$\mathcal{G}_R^n = \begin{cases} i(-R)^m & \text{if } m = nk \\ j(-R)^{m-1} & \text{if } m = nk+1 \\ \dots \\ k(-R)^{m-n+1} & \text{if } m = nk+n-1 \end{cases}. \tag{30}$$

should be considered as the base of $n$-dimensional vectors coding. Thus, the following theorems are true.

**Theorem 7a**. If in $n$-dimensional Euclidian space with base $\{E_i\}$ an algebra is determined, then any point $Z$ of this space may be represented in the positional number system $< \rho = -R, \ A_{R^n} >$, where the set $A$ consists of the vectors (22), and the numbers $\alpha_m \in B_R$.

In particular, for complex numbers there exists a system $< \rho = -R, \ A_{R^2} >$, for example, a quaternary system $<-2, \{0,1,j,1+j\}>$, and for three-dimensional vectors with orts **i, j, k** – an octal system, where each digit takes values (23).

**Theorem 8a**. If in $n$-dimensional Euclidian space with base $\{E_i\}$ an algebra is determined, then any point $Z$ may be represented in a normal positional system

$$< \rho = \pm E_2 \sqrt[n]{R}, \ B_R >. \tag{31}$$

Specifically, for $R=2$ we have a binary system of vector coding by radix (26). For complex numbers there exist systems $< \pm j\sqrt{R}, \ B_R >$, for example, binary system $< \pm j\sqrt{2}, \ \{0, 1\} >$, and for three-dimensional vectors with orts **i, j, k** – the exists a binary system $< \pm j\sqrt[3]{2}, \ \{0, 1\} >$. In the last system we have:

    <i>=1, <-i>=1001, <2i>=1001000, <-2i>=1000;
    <j>=10, <-j>=10010, <2j>=10010000, <-2j>=10000;
    <k>=100, <-k>=100100, <2k>=100100000, <-2k>=100000.

For three-dimensional vectors with orts **i, j, k** there also exists a quarternary number system $< \pm j\sqrt[3]{4}, \{0,1,2,3\} >$, where <4i>=1003000 and <-4i>=1000.

Obviously, for the systems from theorems 7a and 8a the condition (14) is fulfilled.

# References

1. D.E. Knuth. "An Imaginary Number System", Communications of ACM 3, No. 4, 1960.
2. S. Khmelnik. "A Specialized Computer for Operations on Complex Numbers". Questions of Radio Electronics XII, No. 2, 1962 (in Russian).
3. S. Khmelnik. "Adder of codes of complex numbers". Questions of Radio Electronics XII, No. 3, 1965.
4. Penney W., "A 'Binary' System for Complex Numbers", Journal of ACM 12, No. 2, 1965, pp. 247-248.
5. S. Khmelnik. "Positional Coding Systems for Complex Number Presentation". Questions of Radio Electronics XII, No. 9, 1966 (in Russian).
6. S. Khmelnik. "Number system with complex radixes", Part in Book: Pospelov D.A., Arithmetic basis of digital Computers, "Visshaja shkola", 1970, Moscow.
7. S. Khmelnik. "Solution of the navigation problems on a digital computer using complex numbers codes", "Problem special radio-electronic", series "Tele-mechanics and management Systems", 1971, No 6, Moscow.
8. S. Khmelnik et al. "Digital devices with microchips", "Energia", 1975, Moscow.
9. S. Khmelnik. "Computer Arithmetic of Vectors, Figures and Functions", Mathematics in Computers, 1995, Tel Aviv (in Russian).
10. Miller J. M., "Elementary Functions. Algorithms and Implementation". Birkhauser, 1997, Boston.
11. T. Aoki, H. Amada, and T. Higuchi: "Real/Complex Reconfigurable Arithmetic Using Redundant Complex Number Systems". In Proc. 13th Symposium on Computer Arithmetic, 1997.

12. Y. Chang and K. Parhi. "High Performance Digit Serial Complex Number Multiplier-Accumulator", Proc. Int. Conf. on Computer Design,1998.
13. A. M. Nielsen and P. Kornerup, "Redundant Radix Representation of Rings", IEEE Transactions on Computers, Vol. 48 (11), November 1999
14. S. Khmelnik, "A Method and System for Processing Complex Numbers". International Patent Application, WO 01/50332, 2001. . United States Patent Application No. 10/189,195, 2002.
15. S. Khmelnik, "A Method and System for Implementing Coprocessor". Canadian Patent Application, CA 02293953, 2000.
16. S. Khmelnik. "Method and System for Processing Matrices of Complex Numbers and FFT", Canadian Patent Application, CA 2339919, 2001.
17. S. Khmelnik, "Method and System for Processing Matrices of Complex Numbers and Complex Fast Fourier Transformation". International Patent Application, PCT/CA02/00295, 2002
18. S. Khmelnik, A Method and System for Processing Complex Numbers. European Patent Office, EP 1248993. Priority 12.07.01.
19. S.I. Khmelnik, I.S. Doubson, S.M. Khmelnik, A.E. Viduetsky. Arithmetic Unit for Codes by Negative Radix. Publisher «Mathematics in Computers», Israel, Printed in USA, Lulu Inc. No 93352, 146 p., 2004 (in Russian).
20. S.I. Khmelnik. Computer Arithmetic of Complex Numbers and Vectors. Theory, Hardware, Modelling. Publisher «Mathematics in Computers», Israel, Printed in USA, Lulu Inc. No 560836, 387 p., 2006 (in Russian).
21. S.I. Khmelnik, I.S. Doubson, S.M. Khmelnik, A.E. Viduetsky. Arithmetic Unit of Complex Numbers. Publisher «Mathematics in Computers», Israel, Printed in USA, Lulu Inc. No 641090, 110 p., 2007 (in Russian).

# Series: Construction

Evgeny B. Benenson

# A Novel Heat Pipeline Design for North and Northwest Continental Urban and Rural Areas

## Abstract

This article discusses the state of urban and rural heat pipeline networks of various climatic regions. An emphasis is made on vulnerability of the current heat pipeline design, and necessity for some stretch design is being proved. Novel design of various heat pipeline elements is given a fairly detailed consideration; several three-dimensional views of the said elements are also shown. Examples of other countries' heat pipeline structures are given.

The design of the prefabricated elements of the project brought to discussion is not given in full detail. A complete version of the standard-format project (containing parts and units thereof) is available upon potential customers' request. It is important to note that in addition to hot water supply, that is, heat pipeline's routine application, the project is specific for its substantial benefits: as related heat-engineering calculation shows, superficial heat loss in the pipeline under discussion is only 23 wt/sq m, 5 times smaller a figure in contrast to that of conventional structures (119 wt/sq m) currently in use. In terms of fuel equivalent (within the price scale of 1999-2000-ies) this enables to cut expenses for per-kilometer extra heat boosting of return network water by $2,000-3,000 annually. Besides, 70-year-long trouble-free operation period is to be taken into consideration, too. An unbiased evaluation followed by a positive resolution upon the present project by various administrative and governmental bodies or private contractors gives way to the following **opportunities:**

**1.** Development of high-tech production facilities or renovation of the operating ones to manufacture block units and components for efficient heat conduits;

**2**. Availability of governmental or private investments as operation of the high-tech heat conduits is a definitely profitable business (see item 4 below);

**3**. New jobs as a consequence of virtually limitless scope of work to do;

**4**. Unprecedented substantial benefits from operating and controlling the high-tech heat conduit networks involving virtually no overhaul-related expenses whatsoever for a long time;

**5**. A rapid and efficient invigoration of the municipal engineering as well as the entire socio-economical situation in a city or a village.

# Contents

# A. State of the Art.

The underlined urban underground hot-water and heat supply pipelines currently in use seem to be unable to provide reliable long-term operation under minimum heat loss even though built in strict compliance with construction rules and specifications. This can be explained by the fact that simplest possible engineering solutions were applied in heat conduit construction predetermined by shortage of facilities, materials, and time. Besides, no other option seems to have been known by then. While constructing a heat piping climatic conditions, frost penetration, soil mobility, and wandering subterranean waters of the north and northwest regions have not been given due regard. These prerequisites taken together come as cause of emergency conditions and shorter service life. It is only a year's time to observe numerous spots showing complete destruction of the insulation blanket, threatening pipe corrosion, and lower supports flooded with ground water.

Under the standing rules and standards for heat conduits there is no way for altering the situation. Expansion of emergency job scope will remain as endless trouble on an yearly basis. What is more, related expenditures will only be increasing in the course of time.

The western countries employ a tube-in-tube design of the heat conduits where cylindrical inter-tube space is filled with foamed polyurethane to form a heat insulating layer upon solidification. Both central and external tubes are made of a polymer. The design is attractive for its apparent simplicity; there is no need of any slide leg at all, and the tubes do not corrode. This design is given a large-scale dissemination. However it should be noted that the above design is only applicable in the moderate climate zones preferably in horizontal and straight sections of the piping. It should also be remembered that polymer tube connection (lineup and gluing) technique is not easy requiring strict abidance by specifications whatever the conditions. No displacement of the tube edge resulting in cracking is at all admissible. Meanwhile the tubes proper are exposed to the threat of crack-forming at the temperatures ranging within -15 -20°C.

With regard for urban heat supply network configuration complexity, harsh climatic conditions involving sharp temperature differences in the northwest regions and deep frost penetration in the extreme north areas, absolute reliability of the polymer-based heat piping under the above conditions is arguable.

There is a trend of developing independent heat supply systems. *Both efficacy and usability of these systems are irrefutable, however their advent and full-fledged development along with complying with relevant ecological requirements are only possible under specific territorial and economic* **conditions which are still be to be established**. *This is the matter of time.*

Besides it stands to reason to state that through 21-22 centuries and on, the central heating systems will retain their usability.

Developing a novel heat conduit design is an absolute must to radically reduce urban and rural heat conduit maintenance costs **to completely exclude the maintenance or repairs for a long period of operation**, reduce expenditures on fuel of every kind, and finally, save time in the course of economic development.

However, to these ends, *construction concept* for both heat conduit components and urban heat piping is to be radically changed (see *Description of Design and Materials Properties...*).

For the time being, a production project is available to describe now the 2^nd version of the urban heat conduits' components for harsh climatic conditions.

## B. Offer Details

An alternative design of the conduits with pipes ranging 76-325mm in outer diameter is offered **to enable the following**:

1. A complete **protection (confinement) of the space** of hot water conduit from penetration of hostile environments (ground water and the like) using a specially designed concrete shell.
2. Minimum heat loss **(23 wt/sq m)** and heat conduit's external temperature of **8,3 - 9°C** under the heat carrier temperature as high as **100 - 120°C** and environmental temperature of **0°C** (cf.: conventional heat conduit's external temperature is **29°C** (provided the latter is in good working order); and heat loss is as big as **116 wt/sq m** under the same conditions).
3. Excluding a possibility of heat conduit freezing following a sharp temperature decrease or a disconnection from a hot water supply in winter (or because of an emergency).
4. A pre-designed urban heat conduits' ability of faultless operation for a minimum spell of **70 years**. The said long-term trouble-free operation is primarily dependent on both engineering and mechanical state of the heat-carrying tubes as well as quality of welded joints which can be monitored when connection of the components and units is complete.
5. Excluding repairs in the coming years of heat conduit operation. Here, an operating heat conduit of **219–273-mm** in diameter is capable of reducing per-annum fuel expenditures by at least **$2000 for every 1,000 m** of its length (the figures are given for one single-direction heat conduit in the price scale of 1999–2000-ies).

## C. Description of Design and Materials Properties

Along with application of fundamental laws of thermodynamics, calculation methods of heat flow in a closed annular space appearing in formulae proposed by Eckert and academician A.M. Mikheyev (free-convection closed-space dissipation [1]) were used in developing the project.

Structurally, the design of every components of the heat conduit — tees, off-takes (bended tube sections), collectors, and straight sections — is virtually the same. Figures 1, 2, 3, 4 below give a 3D presentation of *some* of the heat conduit's components.

The present heart conduit's design is specific in that the heat-carrying tube proper (hot water tube) is not covered with an insulating layer (see Figures 1, 2, 3, 4).

A cast concrete sheath specific for high moisture and crack resistance, freeze-thaw durability and enhanced strength (1.5-1.65 times as much as compared to conventional structures) comes as main carrying and protective structure. Composition of the concrete sheath includes C3 plasticizers or complex modifying agents; here, consumption of cement is reduced by 20%, according to manufacturer's data. Internal cylindrical surface of the sheath is to be smooth with no coarse or even average grade of finish. Some (special) conditions enable to use an external surface of 2-3mm-thick plastic tube as a base surface in casting of the concrete sheath. When both forming and normal hardening of the concrete is complete, the sheath has the following specifications: water tightness: W14; ductility: Π5; freeze-thaw durability: P400; strength: 53 MPa (at the age of 29 days) and 37 MPa (at the age of 7 days). The sheath's outer surface can be of cylindrical shape, while the inner surface can be square or square-unit one (for 2, 3, or 4 pipelines).

The sheath's inner surface is covered with a cylindrically-shaped thermal insulation layer with thickness corresponding to a specific thermal design (60-50 mm in the selected version). Fiberglass having heat conductivity factor of 0,035 wt/(M*°C) or a *foam-insulating plastic* having heat conductivity factor of 0,035 – 0,04 wt/(M*°C), are used as heat insulators. The heat-insulating layer is formed consecutively form separate cylinder-shaped units 200, 300, 500mm in length (depending on specific nominal size or format). The butt ends are cone-shaped $(90°)$ to ensure tighter coupling. Other reasonable versions like filling preliminarily limited cylindrical space with a thick foamed polyurethane 0,044 wt/(M*°C), are definitely welcome.

A steel tube with anticorrosion coating on both inner and outer surfaces is positioned centrally. The tube is kept in its central position by means of several special slide legs (2 -3 pcs. per a 8-9-meter-long straight section); the number of legs is defined according to specific design and related calculations. These are integrally cast legs made of a heat-resistant polymer containing a metal insert from the side of sliding surface. This polymer component withstands heat exposure of 90-140°C; material's estimated allowable stress is within 80-160MPa; density is 0.92-1.78 g/cub cm. The legs rest on the inner surface of the concrete sheath (see Fig. 4), and are mounted in a set order jointly with installation of the heat-insulation units.

Some air space serving as additional heat resistance factor is formed between external surface of the steel tube and inner surface of the insulation layer as a result of assembling. Layer's effective height is to be defined through a corresponding thermal calculation.

Every thermal broadening of the steel tubes within the sheath with regard for thermal variators' design is duly taken into account.

Installation of specially designed steel expansion pipes is envisaged at the junction points of old and new pipelines to prevent moisture from stealing into new heat conduit's environment.

The legs (see Fig. 1) positioned at the site of pipe-laying ensure general requirements as to pipeline's proper positioning. The legs' design adds to easier and faster components' reciprocal positioning during on-site connection. Specially designed mini-sized cylindrical piles coming as part of the legs' design are oriented perpendicular to the legs' horizontal axial plane to prevent ant displacement or skewing in other crossing planes of the heat conduit's central axis thereby improving heat conduit's overall stability.

Other specific designs of the heat conduits can also be made. However, formation of closed circular space around the heat-carrying tube, that is, the core principle of the present concept is to be left unchanged.

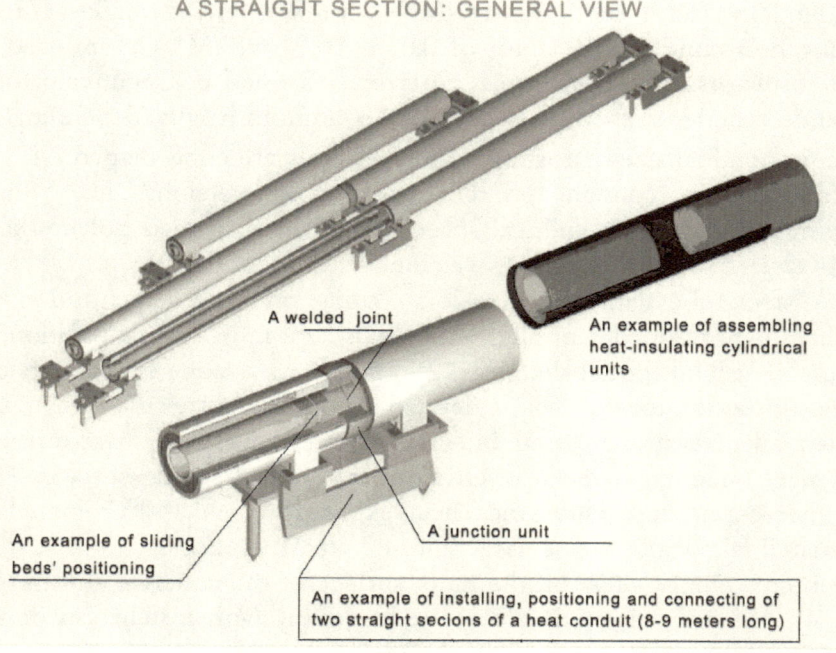

A STRAIGHT SECTION: GENERAL VIEW

A welded joint

An example of assembling heat-insulating cylindrical units

An example of sliding beds' positioning

A junction unit

An example of installing, positioning and connecting of two straight secions of a heat conduit (8-9 meters long)

Fig. 1

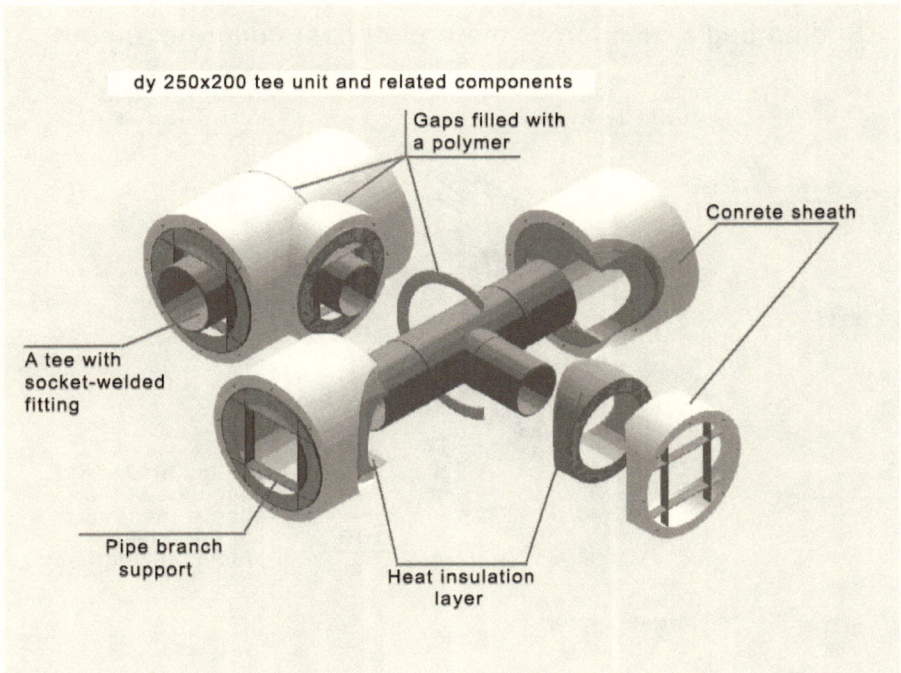

Fig. 2

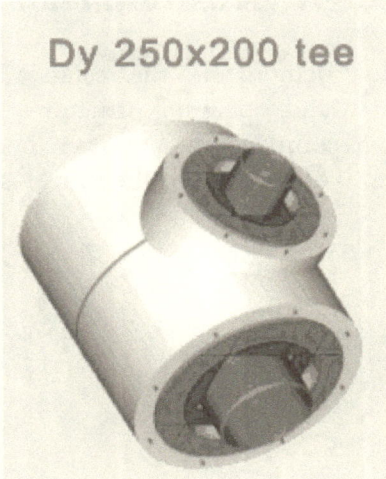

Fig. 3

## Sliding bed's position at the end of cast concrete sheath

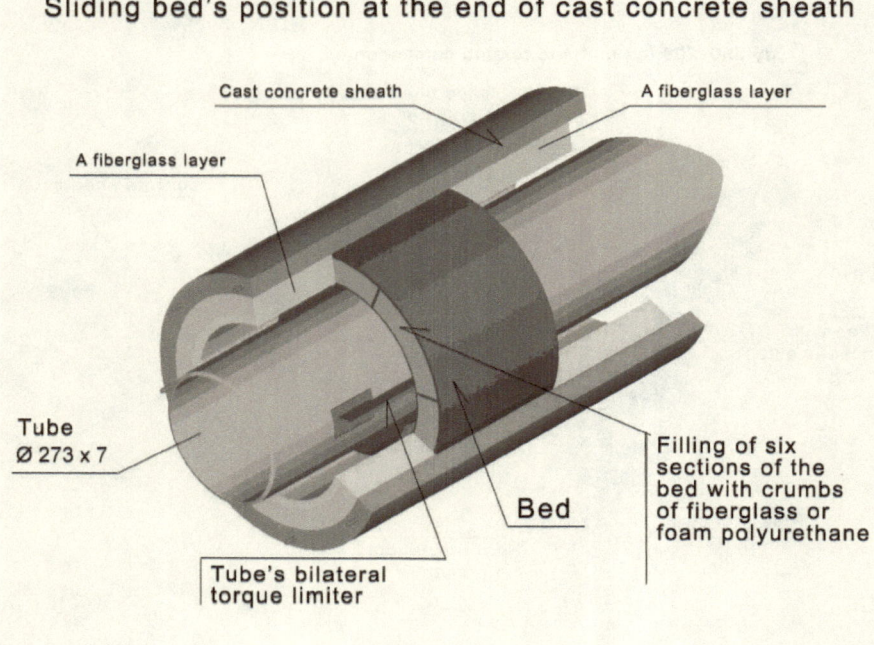

Fig. 4

# D. Engineering Calculation Data

Charts shown below demonstrate the results of thermal engineering calculations of the new and operating structures of underground hot water conduit. Here, a comparative minimum heat loss value in two different structures and efficiency of the novel structure offered are shown.

## Charts of comparative temperature dependencies and superficial thermal losses in two types of heat conduits under ambient temperature variation

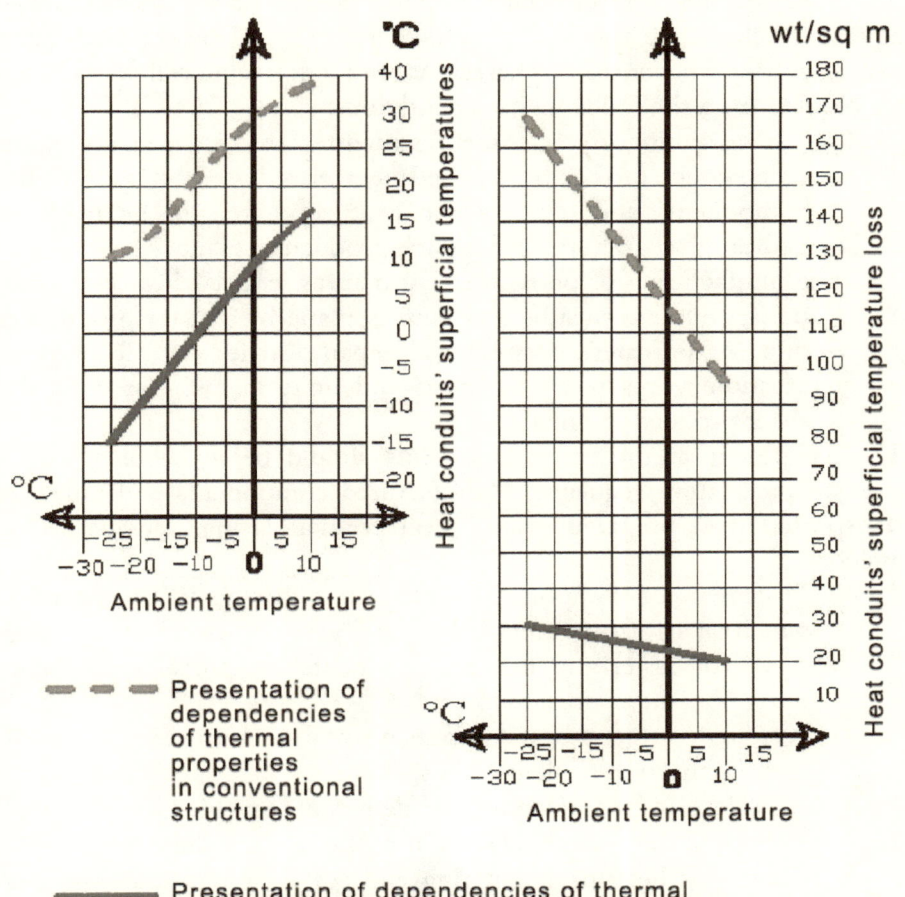

Presentation of dependencies of thermal properties in heat conduits of novel design

# F. Terms of Embodiment

All parts, elements and units of the hear conduits are to be manufactured in plant's conditions. To these ends, development of dedicated production facilities is necessary. There should be no place to a doubt of a possibility of having the said facilities equipped correspondingly. There is related experience in developing and manufacturing of cylinder-shaped heat-insulating units complying with the present offer's specifications. There are related technology and experience of manufacturing the cast concrete sheaths corresponding to the present offer's specifications. And finally, there must be no doubt as to designer's ability for developing appropriate building berths, master plates and machinery to be built for further operation at workshops. The equipment should be completely faultless in terms of heat conduit's parts manufacturing and erection phases. A production unit can be selected proceeding from the following two conditions:

1. Manufacturing of parts, metal structures, circular heat insulation units, concrete sheaths and beds as well as assembling of all the components and units of the heat conduit is to be made at dedicated workshops of the same production plant.

2. Manufacturing of parts, metal structures, circular heat insulation units, concrete sheaths and beds is made by a sister production unit. A dedicated assembling company carries out all kinds of erection works using the parts and units manufactured by the sister production units.

The equipment of the production units should be in keeping with a selected production sequence. Manufactured components and units are transported to a prepared site of heat conduit laying. On-site works include:

1. Placement of components and units onto special setting piles which apart from complying with general specifications also serve for easier and faster components' reciprocal positioning during on-site connection.

2. Welding of steel tubes' components and units to at the point of their connection to sliding beds.

3. Non-destructive ultrasonic flaw detection in welded joints.

4. Testing of the assembled section of the heat conduit.

5. Placing of heat-insulation ring and sealing joints units using a special technique. The technique enables a safe binding of the sealing ring with the bulk of the concrete sheath of components and units eliminating a threat of crack formation at the connection points.

Provided a well-reasoned engineering preparation is done, the rate of heat conduit laying is 3-5 times faster than the one made following conventional construction sequence.

# References

1. B.N. Yudayev. Heat Transfer. «The "Vysshaya Shkola" Publishers», 1973.
2. B.N. Yudayev. Collected Problems in Thermodynamics and Heat Transfer. «The "Vysshaya Shkola" Publishers», 1964.
3. B.V. Rudomino, Yu.N. Remzhin. "Designing Combined Power Plants. «The "Energiya" Publishers», 1970.
4. E.B. Benenson. The new construction of the hot-water systems for the thermal networks of cities and settlements. "The Papers of Independent Authors", publ. «DNA», printed in USA, Lulu Inc. 124173, Israel-Russia, 2005, iss. 1 (in Russian).
5. E.B. Benenson. The new construction of the hot-water systems for thermal networks within the limits of city line and settlements http://laboratory.ru/articl/tech/at211.htm (in Russian).

# Series: **Physics and Astronomy**

**Mihail A. Karpov**

# The spectrum of masses of elementary particles, relationship between micro- and macroscales, relation ship among cosmic energies

## Abstract

Alternate hypothesis for the concepts of the matter structurization is set out. The mass values of some known as well as previously unknown particles have been theoretically obtained on a basis of new results conformities to natural law. This hypothesis provides the means of indegration of world of micropartiles with cosmological concepts as dark energy and dark matter.

## Contents

## Introduction

Until very recently, it was agreed that the building of contemporary physics had already essentially been erected. The only some additions are to be made and the further advancement of physics is deadlock. However, a body of recent astronomical investigations has uncovered that the picture of the Universe has been something of a mystery or in more exact terms, is not satisfied earlier notions. These astronomical investigation results are, above all, accelerated expansion of negative pressure of vacuum or dark energy ("quintessence"). Besides, there is invisible dark matter surrounding galaxies. The constitution of this matter is unknown. The common baryon substance forming the stars is only small part of the Universe mass.

In addition, there is "something" in Solar system. This "something" retarded the motion of cosmic sondes "Pioneer", launched at the 1970s and gave no way to escape the limits of Solar system.

Together these and many others facts cause us to take a fresh look at eternal questions: what is the matter and what is the space? What is manner in which the world of microparticles and macroscales are related to each other? What is the association between gravitation and other interactions? It is obvious that such relationship is to be. This article is concerned with abovementioned questions.

## 1. The elementary charge and the spectrum of masses of elementary particles or why the μ-meson is 207 times heavier than electron

The nature surrounding us is unified and laws controlling the world of microparticles are inextricably entwined with the laws of shaping and progression of matter macrostructures and the Universe as a whole. The gravitation as physical field and inertia are the base for the shaping of the all interactions of forces and charges of macro- and microworld.

The electron is fundamental and stable particle having elementary elektrical charge. Think of a electron as a "system" containing the virtual Planck particle of spin $h$ with dimension $r_{pl} = \left(\dfrac{Gh}{c^3}\right)^{1/2}$ , and virtual particle $m$, radially "moving" in scale $r$ in the rotating field of gravitation of the Planck particle.

Let us assume that the energy of electrical charge is

$$\frac{e^2}{r} = \frac{h^2}{mr^2} \cdot \frac{r_{pl}^2}{r^2}. \tag{1}$$

This relation may be written as:

$$r = \frac{h^2}{me^2} \cdot \frac{r_{pl}^2}{r^2}$$

and now the relation is somewhat similar to the Bohr quantum conditions:

$$r_n = \frac{h^2 \cdot n^2}{m_e \cdot e^2}.$$

Considering the value of $r_{pl}$, the r would be expressible as:

$$r = \frac{h}{c}\left(\frac{G}{me^2}\right)^{1/3}. \tag{2}$$

Assigning to the masses of virtual particles $m$ the quantum numbers $m_n = m_0 \cdot e^{-n^2}$, where $m_0$ is the mass of most heavy particle (subsequent to $m_{pl}$) and $n = 0, 1, 2,\ldots, 12$, we derives

$$r_n = \frac{h}{c}\left(\frac{G}{m_0 \cdot e^{-n^2} \cdot e^2}\right)^{1/3}. \tag{3}$$

Table 1

| $n$ | 1 | 2 | 3 | 4 |
|---|---|---|---|---|
| $m_{obs}(m_e)$ | 1.4 | 3.8 | 20 | 207 |
| $m_{vir}(m_e)$ | $2.6 \cdot 10^{20}$ | $1.3 \cdot 10^{19}$ | $8.6 \cdot 10^{16}$ | $7.8 \cdot 10^{13}$ |
| $l_{vir}(cm)$ | $10^{-31}$ | $10^{-30}$ | $10^{-27}$ | $10^{-24}$ |
| $2r_n(cm)$ | $2 \cdot 10^{-31}$ | $5 \cdot 10^{-31}$ | $2 \cdot 10^{-30}$ | $2 \cdot 10^{-29}$ |

| $n$ | 5 | 6 | 7 | 8 |
|---|---|---|---|---|
| $m_{obs}(m_e)$ | 4160 | 162700 | $1.2 \cdot 10^7$ | $1.8 \cdot 10^9$ |
| $m_{vir}(m_e)$ | $10^{10}$ | 162700 | $3.6 \cdot 10^{-1}$ | $1.1 \cdot 10^{-7}$ |
| $l_{vir}(cm)$ | $10^{-20}$ | $10^{-16}$ | $10^{-10}$ | $10^{-4}$ |
| $2r_n(cm)$ | $4 \cdot 10^{-28}$ | $2 \cdot 10^{-26}$ | $3 \cdot 10^{-24}$ | $4.5 \cdot 10^{-22}$ |

| $n$ | 9 | 10 | 11 | 12 |
|---|---|---|---|---|
| $m_{obs}(m_e)$ | $5.3 \cdot 10^{11}$ | $3 \cdot 10^{14}$ | $3.3 \cdot 10^{17}$ | $7 \cdot 10^{20}$ |
| $m_{vir}(m_e)$ | $4.6 \cdot 10^{-15}$ | $2.6 \cdot 10^{-23}$ | $2 \cdot 10^{-32}$ | $2 \cdot 10^{-42}$ |
| $l_{vir}(cm)$ | $10^4$ | $10^{12} \div 10^{13}$ | $10^{21} \div 10^{22}$ | $10^{31}$ |
| $2r_n(cm)$ | $1.5 \cdot 10^{-19}$ | $10^{-16}$ | $10^{-13}$ | $2.4 \cdot 10^{-}_{10}$ |

It is suggested that the observed mass of particle is bound with this dimension by simple relation

$$m_{obs} = kr,$$

where $k$ is constant.

Assume that the particle is electron at $n = 0$, $m_{vir} = m_0$. Its dimension is $r_0 \sim 10^{-31}$ cm (do not confuse with the classical election radius). Inserting $n = 0, 1, 2, \ldots, 12$ in Eq.3, we arrive the dimensionless relations or $\dfrac{m_{obs\,n}}{m_e}$, as mentioned above. Arrange the arrived relations in tabulated form (Table 1, first line).

By this means the masses of observed particles increase up to
$$7 \cdot 10^{20}\, m_e \sim 3.5 \cdot 10^{17} \text{ GeV } (n = 12).$$

When $n = 13$, mass of particles are greater than mass of Planck $(1.2 \cdot 10^{19} \text{ GeV})$.

On the assumption that this mass is $m_0$ in Eq. 3, we calculate the second line of Table. The dimension of virtual particle (wave length), compatible with its mass is in third line of Table. In fourth line of Table we calculate the "doubled" radius (3). When $n$ is major the wave lengths of virtual particles are enormous. However, it should be remembered that in the "constitution" of observed particles these virtual particles are in the "bound" and not free state.

Once again, we call attention to the observed particles

$$\frac{r_1}{r_0} = 1.4, \qquad \text{at } n = 1, \qquad \frac{r_4}{r_0} = 207, \qquad \text{at } n = 4,$$

$$\frac{r_2}{r_0} = 3.8, \qquad \text{at } n = 2, \qquad \frac{r_5}{r_0} = 4160, \qquad \text{at } n = 5,$$

$$\frac{r_3}{r_0} = 20, \qquad \text{at } n = 3, \qquad \frac{r_6}{r_0} = 162700, \qquad \text{at } n = 6.$$

When $n = 4, 5, 6$ these relations somewhat resemble, with known error, masses of $\mu$- meson, $\tau$- meson and $w$- boson expressed in terms of the masses of electron.

The mass of actual $\tau$- lepton is less than mass of (4160) state by value of mass (or binding energy which is equal to this mass) of two virtual $\pi$- mesons which are the disintegration product of $\tau$- lepton in parallel with $\nu_\tau$. Consequently, the part of energy of this state is used for the binding energy of two $\pi$- mesons.

Hadrons or strong interaction particles may be expressed as derivates of $\mu$, $\tau$ and $w$ states:

$$\mu\text{ - hadron} \qquad \tau\text{ - hadron} \qquad w - \text{hadron}$$

$$\begin{pmatrix} \pi, p, n \\ u, d \end{pmatrix}\begin{pmatrix} K \\ S \end{pmatrix} \qquad \begin{pmatrix} J/\Psi \\ d \end{pmatrix}\begin{pmatrix} \gamma \\ S \end{pmatrix} \qquad \begin{pmatrix} t\text{ - quarks} \\ S \end{pmatrix}$$

where $d(1.4)$, $u(3.8)$, $S(1.4 \times 3.8)$ are levels.

Table 2

| $n$ | 0 | 1 | 2 | 3 | 4 | 5 | 6 | Not |
|---|---|---|---|---|---|---|---|---|
| | $e$ | $d$ | $u$ | $c$ | $\mu$ | $\tau$ | $w$ | e |
| $e^-$ | + | - | - | - | - | - | - | |
| $\mu^-$ | + | - | - | - | + | - | - | |
| $\tau^-$ | + | - | - | - | - | + | - | |
| $\pi^-$ | + | + | - | - | + | - | - | ↑↓ S |
| | - | - | + | - | + | - | - | |
| $p^+$ | + | + | - | - | + | - | - | d |
| | - | - | + | - | + | - | - | u ↑↓↑ S |
| | - | - | + | - | + | - | - | u |
| $n^0$ | + | + | - | - | + | - | - | d |
| | + | + | - | - | + | - | - | d ↑↓↑ S |
| | - | - | + | - | + | - | - | u |
| $K^-$ | + | + | + | - | + | - | - | ↑↓ S |
| | - | - | + | - | + | - | - | |
| $J/\Psi$ | + | + | - | - | - | + | - | $c\bar{c}$ |
| | + | + | - | - | - | + | - | |
| $\gamma$ | + | + | + | - | - | + | - | $b\bar{b}$ |
| | + | + | + | - | - | + | - | |
| $W^-$ | + | - | - | - | - | - | + | ↑↑ S |
| | - | - | - | - | - | - | + | |
| $z^0$ | + | - | - | - | - | - | + | ↑↑ S |
| | + | - | - | - | - | - | + | |
| $t\bar{t}$ | + | + | + | - | - | - | + | $t\bar{t}$ |
| | + | + | + | - | - | - | + | |

The masses of $\gamma$-meson and $J/\Psi$-meson are in the ratio as masses of $K$ and $\pi$-mesons.

$$\frac{m_\gamma}{m_{J/\Psi}} = \frac{11020\ \text{MeV}}{3100\ \text{MeV}} \approx \frac{m_K}{m_\pi} \cong 3.5 .$$

Likewise the masses of associated $b$ and $c$-quarks are in the ratio:

$$\frac{4.6\,\text{GeV}}{1.3\,\text{GeV}} \cong 3.5 .$$

In such a manner, it has been proposed that $J/\Psi$ particle (or $c\bar{c}$) is $\tau$ state of $\pi$-meson and $\gamma$ particle (or $b\bar{b}$) is $\tau$ state of $K$- meson. $D$ and $B$-mesons are the combinations of corresponding $c$ and $b$ quarks with quarks of $\mu$ - state.

The Table 1 is shown that mass of virtual particle is equal to mass of observed particle at $n = 6$. The mass of virtual particle at $n = 0$ is equal to mass of observed particle at $n = 12$ ($7 \cdot 10^{20}\, m_e$). These particles ($n = 6$, $n = 12$) are $W$ and $X$ – bosons respectively. $t$-quark is $W$- state of $K$- meson. $W$ – state of $\pi$- meson and state with $n = 3$ are not realized as particles.

It should be noted that the ratio of the masses of existing hadrons to the masses of quarks is identical order as the ratio of masses of $\mu$, $\tau$ and $w$ states

$$\frac{m_\gamma}{m_k} \sim \frac{m_{J/\Psi}}{m_\pi} \sim \frac{m_\tau}{m_\mu} \sim 20; \quad \frac{m_t}{m_b} \sim \frac{m_W}{m_\tau} \sim 40.$$

According to quantum states the elementary particles may be shown in tabulated form (see Table 2). It is evident from the Table that the particle is the combination of two or more states (with the exeption of electron). The particles with one or two quantum levels are interacting only rarely.

The particles with three levels and up are hadrons.

The birth of massive lepton with $n = 7$ and mass about 6 Tev is likely with on further increasing the energy of colliders.

Once again, we call attention to the Table 1. As supposed above, a electron is structure containing the virtual particle of mass $m_0$ (heaviest after of $m_{pl}$, $n = 12$, $m = 7 \cdot 10^{20}\, m_e$), «moving» in radius $r_0 \sim 10^{-31}$ cm. The particle $m_0$ itself is made up by the «cloud» from lightiest particle of mass $2 \cdot 10^{-42} m_e$, «moving» in radius $r_{12}$.

Now work out the $2r_{12}$ or Compton length for the electron:

$$2r_{12} = \frac{2h}{c}\left(\frac{G}{m_0 \cdot e^{-144} \cdot e^2}\right)^{1/3} \quad (n = 12),$$

(4)

where $m_0 = m_e \dfrac{r_{12}}{r_0} = m_e \cdot e^{\frac{144}{3}} \cong 7 \cdot 10^{20} m_e$.

Alternatively, Compton length for the electron is equal to

$$\lambda_e = \frac{h}{m_e c} = 2r_{12}.$$

Equating this value to Eq. 4, we obtain:

$$m_e^3 = \frac{m_e \cdot e^{\frac{144}{3}} \cdot e^{-144} \cdot e^2}{8G} \Rightarrow$$

$$\Rightarrow m_e = \frac{e \cdot \exp\left(-\frac{1}{3} \cdot 144\right)}{\sqrt{8G}} = m_0 \cdot \exp\left(-\frac{144}{3}\right),$$

where $m_0 = \dfrac{e}{\sqrt{8G}}$ is magnetic monopole ($e$ is electron charge, $G$ is gravity constant).

As a result we obtain:

$$2r_{12} = \frac{h}{m_e c} = \frac{h}{c} \cdot \frac{\sqrt{8G}}{e} \cdot \exp\left(\frac{144}{3}\right) \cong 2.4 \cdot 10^{-10} \, \text{cm}.$$

Consequently, $2r_{12} \cong 2.4 \cdot 10^{-10}$ cm at $n = 12$, this value is correlated with Compton length for the electron.

We obtain $2r_{11} \cong 1.2 \cdot 10^{-13}$ cm at $n = 11$, this value is correlated with characteristic of the range of strong interaction.

The heavy particles mass whereby the "Great" union, as it is called, occurs and, which controls the life time of proton is $3.3 \cdot 10^{17}$ $m_e$ or $10^{14} \div 10^{15}$ GeV.

$2r_{10} \cong 10^{-16}$ cm at $n = 10$ we obtain the characteristic of the range of slight interaction.

Consequently, the mass $\dfrac{r_{12}}{r_0}$ of heaviest particle $m_0$ or $\chi$ - particle is $7 \cdot 10^{20}$ $m_e$. This mass is above the mass of "Great" union. The mass or lightest particle is $2 \cdot 10^{-42}$ $m_e$. Hence, the "dynamic range" of mass spectrum is $3.5 \cdot 10^{62} = e^{144}$. We designate this "dynamic range" as constant $\alpha$ or **constant of inflation**. We shall cover this constant later.

For the time being, we point out that the Eq. (1), for macroscale (e.g. the Earth), will be occur as:

$$\frac{Q^2}{r} = m \cdot \frac{V_{es}^2}{c^2} \cdot \dot{r}^2 \cdot \frac{\omega^2 r}{g_0} \cdot \frac{g(r)}{g_0},$$

where, $Q$ is induced charge, $m$ – the body mass having radial component of velocity $\dot{r}$ in the field rotating with frequency $\omega$ in gravity field of massive body with potential $V_{es}^2 = \dfrac{GM}{r}$ - escape velocity squared, $g_0$ – the acceleration of gravity on surface of massive rotating body. When substituting $\quad \omega = \dfrac{h}{mr^2}, \quad V^2{}_{es} = \dfrac{m_{pl} G}{r}, \quad g_0 = \dfrac{m_{pl} G}{r^2}, \quad \dfrac{g(r)}{g_0} = \dfrac{r_{pl}^2}{r^2},$

$Q = e$, $\dot{r} = c$ we obtain the Eq. (1). It seems likely that this mechanism is universal and produces the electromagnetic effects which arise at radial motion of bodies in the rotating gravity field of massive body.

In the integrated form, this is expressible as:

$$\frac{\partial}{\partial t}\frac{\varphi}{c^2} \cdot \int_V [\vec{r}\,\vec{v}]\rho\, dV = \oint [\vec{E}\,\vec{H}]dS \quad \text{or} \quad \frac{\varphi}{c^2} \cdot \frac{\partial L}{\partial t} = W_{em},$$

where $\varphi = \dfrac{GM}{r}$ is potential, $L$- torque, $W$ – electromagnetic energy.

By this is meant that the variation in the torque, inside the surface $S$, implies a flow of electromagnetic energy through this surface.

## 2. On the association of macroscales and microscales and feasible scenario of the progress of the Universe

1. If we denote the abovementioned value $e^{144} \cong 3.5 \cdot 10^{62}$ by $\alpha$, then the association between the radius of the Universe and the Planck radius is:

$$\alpha \cdot l_{pl} = R_U \cong 3 \cdot 10^{29} \text{ cm},$$

and between the mass of the Universe and Planck mass is

$$\alpha \cdot m_{pl} = M_U \cong 10^{57} \text{ g}.$$

2. It the maximum density of substance is $\dfrac{m_{pl}}{l_{pl}^3}$ (that is the Planck mass per the quantum of space)and if we suppose that the minimum density of substance is $\dfrac{m_{pl}}{\left(\alpha \cdot l_{pl}\right)^3}$, that, on other hand, is the mass quantum per the all "range" of space, we obtain:

$$\rho_{min} \cong \frac{10^{-5} \text{ g}}{10^{88} \text{ cm}^3} \cong 10^{-93} \text{ g/cm}^3; \quad R_U^3 = 10^{88} \text{ cm}^3,$$

in which case the Universe mass is:

$$\frac{\alpha \cdot m_{pl}}{\alpha \cdot R_U^3} = \frac{M_U}{\alpha \cdot R_U^3} = \rho_{min}; \quad \alpha \cdot R_U^3 = R_{cr}^3$$

and critical radius is

$$R_{cr} = \alpha^{1/3} \cdot R_U \cong 10^{50} \text{ cm} = \alpha^{4/3} \cdot l_{pl}.$$

This value is in close agreement with the life time of proton multiplied by the velocity of light:

$$R_{cr} = c\tau_p \cong 10^{50} \text{ cm}, \quad \tau_p = 10^{33} \text{ years} \sim 10^{40} s.$$

3. At the instant of the generation (formation) of the Universe, the period of exponential expansion (inflation) – radius and mass (as one of dimensions) of six-dimensional structure (Planck element) increased by $\alpha = e^{144}$ раз. That is

$$R_U = \frac{l_{pl}}{2} \cdot 3.5 \cdot 10^{62} \approx 2.5 \cdot 10^{29} \text{ cm},$$

$$\left( l_{pl} = \sqrt{\frac{G\hbar}{c^3}} \cong 1.5 \cdot 10^{33} \text{ cm} \right).$$

$$M_U = 2 \cdot 10^{-5} \cdot 3.5 \cdot 10^{62} \approx 7 \cdot 10^{57} \text{ g},$$

$$\left( M_{pl} = \sqrt{\frac{c\hbar}{G}} \cong 2 \cdot 10^{-5} \text{ g} \right).$$

At the same time, the sixdimensional sphere of Planck decays into $\chi$-particles ($n=12$) with mass

$$m_\chi = \frac{e}{\sqrt{8G}} \approx 7 \cdot 10^{20} m_e = 3.5 \cdot 10^{17} \, \text{GeV}$$ , that is that competing

on equal terms, six dimensions decay in to substance and space-time.

The ratio between Planck mass and $\chi$- particle mass is

$$\frac{m_{pl}}{m_\chi} = \frac{\sqrt{\dfrac{c\hbar}{G}}}{\dfrac{e}{\sqrt{8G}}} \cong 33.1 = \sqrt{\frac{8}{\alpha_{fs}}} \, ,$$

where $\alpha_{fs} = \dfrac{e^2}{\hbar c} = \dfrac{1}{137}$ is constant of fine structure (not to be con-

fused with inflation constant $\alpha$).

The area of six dimensional sphere in seven dimensional space is

$$S_6 = \frac{16}{15} \pi^3 R^6 \cong 33.07 R^6 \, ,$$

that is that the Planck six dimensional sphere "involves" $\left(\dfrac{m_{pl}}{m_\chi}\right)^6$ $\chi$-

particles (since the mass is one of dimensions).

$$\left(\frac{m_{pl}}{m_\chi}\right)^6 = 1.3 \cdot 10^9 = \beta = \frac{n_\gamma}{n_B}$$

is the fundamental constant of cosmology which are responsible for the baryon asymmetry of the Universe (ratio between amount of photons of relict rays and baryons per unit of volume).

Alternatively, from the experimental data, constant $\beta$ is worked out as:

$$\beta = \frac{n_\gamma}{n_B} = \frac{400}{3 \cdot 10^{-7}} \approx 1.3 \cdot 10^9 \, ,$$

where $n_B$ is density of baryons

$$n_B = \frac{N}{V} = \frac{4.4 \cdot 10^{81}}{1.5 \cdot 10^{88}} \cong 3 \cdot 10^{-7} \, \frac{1}{\text{cm}^3} \, ,$$

it is the present-day evalution from astronomical data.

$$V = (2.5 \cdot 10^{29} \text{ cm})^3 \cong 1.5 \cdot 10^{88} \text{ cm}^3$$

is the Universe volume.

$$N = \frac{7 \cdot 10^{57} \text{ g}}{1.6 \cdot 10^{-24} \text{ g}} = \frac{M_U}{m_p} \cong 4.4 \cdot 10^{81}$$

is amount of baryons ($m_p$ – mass of proton). $n_\gamma \cong 400 \dfrac{1}{\text{cm}^3}$ is density of photons of relict radiation.

That is that the above calculated value is in close agreement with evaluation obtained by experimental methods.

4.  Thus, at first stage of rising of matter, at $n = 12$ and $E \sim 10^{17}$ GeV within the first $10^{-6}$ s (before hadron era), the quark $\leftrightarrow$ lepton transformations take place. These transformations are followed by the rising of lightest particles in amount of:

$$N_{12} = \frac{T_{12}}{\tau_{12}} N\beta = \frac{10^{-6} s}{1.7 \cdot 10^{-42} s} \cdot 1.3 \cdot 10^9 \cdot 4.4 \cdot 10^{81} \cong 3 \cdot 10^{126},$$

where $\tau_{12} = \beta^{1/6} \cdot \tau_{\text{pl}} = 33.1 \cdot \tau_{\text{pl}} \cong 1.7 \cdot 10^{-42} s$ is characteristic time of the rising of these particles.

With decreasing energy to $3.3 \cdot 10^{17} m_e \sim 10^{14}$ GeV (energy of integration of electro-weak and strong interactions $n = 11$ (see Table 1)), the rising-decay of baryons takes place. In the modern view, at ultra-high temperatures and densities the plausible characteristic time of the rising-decay of baryon in quark-baryon plasma is of the order of $10^{-28}$–$10^{-29}$ sec. This process extends about $10^{-4}$ sec since the onset of explosion and up to $10^{13}$ K.

Each of the of rising-decay of baryon are followed by the rise of light particle $n = 11$; $m = 2 \cdot 10^{-32} m_e$; $r = 10^{21} \div 10^{22}$ cm (see Table. 1).

The total amount of these light particle is:

$$N_{11} = N \, \beta \frac{10^{-4} s}{\tau} = 4.4 \cdot 10^{81} \cdot 1.3 \cdot 10^9 \cdot \frac{10^{-4} s}{10^{-29} s} \cong 10^{116}.$$

The factor $\beta$ occurs because the annihilation of antibaryons takes place after the rises of light particles.

Considering that the mass of these particles is more than mass of light particles ($n = 12$, $r = 10^{31}$ cm) by a factor $10^{10}$, we obtain that its total mass exceeds the mass of substance (baryons) of the Universe. It seems likely that it is hidden mass (invisible halo of galaxies).

By analogy, we may assume that at transition of next level ($n = 10$, weak interaction), e.g. at neutralization of stars and flares of ultra new stars, the rise of large amount of light particle ($n = 10$, $r = 10^{12} \div 10^{13}$ cm) takes also place. This process is fovourable to escaping heavy elements and generation of planetary systems.

## 3. On the association of universal constants with the values $\alpha$ and $\beta$

As indicated above, value $\alpha$ is the constant of inflation $e^{144} \cong 3.5 \cdot 10^{62}$. The value $\beta$ or cosmological value defining the degree of baryon asymmetry of the Universe is equal to the ratio between the amount of relict photons and the amount of baryons:

$$\frac{n_\gamma}{n_B} \cong \frac{400 \dfrac{1}{\text{cm}^3}}{3 \cdot 10^{-7} \dfrac{1}{\text{cm}^3}} \cong 1.3 \cdot 10^9 = \beta .$$

From the multidimensional geometry it is known that surface of six-dimensional sphere is equal to $S_6 = \dfrac{16}{15}\pi^3 R^6 \cong 33.07 R^6$, with the dimensionality of space is $n = 7$. The surfaces of five-dimensional sphere and seven-dimensional sphere are accordingly equal to $S_5 = \pi^3 R^5 \cong 31 R^5$, $S_7 = \dfrac{\pi^4}{3} R^7 \cong 32.47 R^7$. Notice that the surface $S_6$ has maximum factor 33.07 for all values $S_n$.

If we suppose that, hand in hand, with inflation, the decay of Planck six-sphere to spheres of radius $33.07 \cdot R$ takes play (that is the rise of substance - $\chi$-protoparticles having mass which is less than Planck mass by factor 33.1), then its amount is equal to: $(33.07)^6 \cong 1.3 \cdot 10^9 = \beta$.

Thus, the $\dfrac{16}{15}\pi^3$ is equal to $\beta^{1/6}$.

The physical values and world constants are measured in centimeter-gramme- second system which is symmetrical to Gaussian system of units wherein, among other factors, $\varepsilon_0 = \mu_0 = 1$. We can assume that the association between these units of measurement of physical values and "large" values $\alpha$ and $\beta$ is available. The "large" values $\alpha$ and $\beta$ define as though the scale (range) of its measurements.

If the three scales-time, density and mass – are sketched in all "range" of values from minimum up to maximum values, we obtain

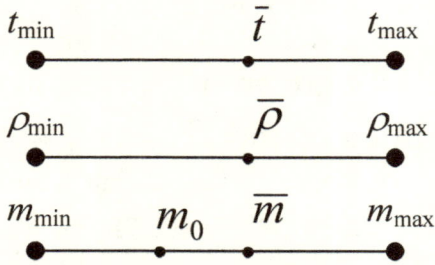

1) $t_{min} = t_0 = t_{pl} \cong 5 \cdot 10^{-44}$ s; $t_{max} = t_0 \cdot \alpha^{4/3} \cdot \beta^{2/6} \cong 2 \cdot 10^{43}$ s, where $t_0 \cdot \alpha^{4/3}$ is life time of nuclons (baryons) ($\sim 10^{40}$ s), and $t_{max}$ is life time of leptons defined by mass of $\chi$- particle $n = 12$.

$$\bar{t} = 1s = \sqrt{t_{min} t_{max}} \approx t_0 \cdot \alpha^{2/3} \cdot \beta^{1/6}.$$

2) $\rho_{min} = \dfrac{m_{pl}}{(\alpha \cdot l_{pl})^3} = \dfrac{\rho_{max}}{\alpha^3} \cong 10^{-94} \text{ g/cm}^3$ ;

$$\rho_{max} = \rho_{pl} = \rho_0 \cong 10^{94} \text{ g/cm},$$

$$\bar{\rho} = \sqrt{\rho_{min}\rho_{max}} = 1 \text{g/cm} \approx \frac{m_0}{l_0^3} \cdot \alpha^{-3/2}.$$

3) $m_0 = m_{pl}$; $m_{min} = \dfrac{m_0}{\beta^{1/6}} \dfrac{1}{\alpha}$ is mass of lightest particle ($n = 12$);

$m_{max} = m_0 \alpha \beta \beta^{1/6}$ is mass of the Universe in "hadron era" before annihilation $(10^{-4}$ s);

$$\bar{m} = 1 \text{g} = \sqrt{m_{min} m_{max}} \approx m_0 \beta^{1/2}.$$

From 2) and 3) we way obtain that $1cm = \left( \dfrac{1\text{g}}{1\text{g/cm}^3} \right)^{1/3}$ and then:

$1 \text{ cm} \approx l_0 \cdot \alpha^{1/2} \cdot \beta^{1/6},$

$1 \text{ g} \approx m_0 \cdot \beta^{1/2},$            (*)

$1 \text{ s} \approx t_0 \cdot \alpha^{2/3} \cdot \beta^{1/6}.$

From these values, we may obtain rough values of the constants:

$$c = \frac{l_0}{t_0} \approx \frac{\alpha^{2/3}}{\alpha^{1/2}}\left(\frac{\text{cm}}{\text{s}}\right) = \alpha^{1/6}\left(\frac{\text{cm}}{\text{s}}\right)$$

- is independent of $\beta$.

$$h = \frac{m_0 l_0^2}{t_0} \approx \frac{1\text{g} \cdot \beta^{-1/2}\alpha^{-1} \cdot \beta^{-2/6}\,\text{cm}^2}{1\text{s} \cdot \alpha^{-2/3} \cdot \beta^{-1/6}} = \alpha^{-1/3} \cdot \beta^{-2/3}\frac{\text{g} \cdot \text{cm}^2}{\text{s}}.$$

$$G = \frac{l_0^3}{m_0 t_0^2} \approx \frac{\beta^{1/2}\alpha^{4/3} \cdot \beta^{2/6}}{\alpha^{3/2} \cdot \beta^{3/6}} = \alpha^{-1/6} \cdot \beta^{1/3}\frac{\text{cm}^3}{\text{g} \cdot \text{s}^2}.$$

The higher approximation of these constants is arrived multiplaying the Eq. (*) in to the constants near 1:

$$\begin{cases} 1cm = l_0 \cdot \alpha^{1/2} \cdot \beta^{1/6} \cdot k_1, & k_1 = \left(\dfrac{S_6}{S_7}\right)^{1/6} = 1.0031, \\[2em] 1g = m_0 \cdot \beta^{1/2} \cdot k_2, & k_2 = \left(\dfrac{4S_6}{S_5}\right)^{1/6} = 1.2375, \\[2em] 1s = t_0 \cdot \alpha^{2/3} \cdot \beta^{1/6} \cdot k_3, & k_3 = \left(\dfrac{2S_6}{S_5}\right)^{1/6} = 1.1346, \end{cases}$$

where $S_5$, $S_6$, $S_7$ – are surfaces of unit spheres in six-, seven- and eight-dimensional space. With these results, we obtain:

$$c = \frac{k_3}{k_1}\alpha^{1/6} = 2.996 \cdot 10^{10} \text{ cm/s},$$

$$h = \frac{k_3}{k_2 k_1^2}\alpha^{-1/3} \cdot \beta^{-2/3} = 1.055 \cdot 10^{-27}\left(\text{g} \cdot cm^2\right)/\text{s},$$

$$G = \frac{k_3^2 k_2}{k_1^3}\alpha^{-1/6} \cdot \beta^{1/3} = 6.68 \cdot 10^{-8}\,\text{cm}^3/\left(\text{g} \cdot s^2\right),$$

where

$$\alpha = e^{144}, \quad \beta^{1/6} = \frac{16}{15}\pi^3.$$

## 4. The alternative description of elementary particles or what is the color of quark

As evident from the Table 1, an electron is particle formed by double motion ($n = 0$, $m_{vir} = 7 \cdot 10^{20} m_e$) and in its turn $m = 7 \cdot 10^{20} m_e$ at $n = 12$ is "formed" by the motion of particle $m_{vir} = 2 \cdot 10^{42} m_e$. The radius of single "inside" motion is equal to $\sim 10^{-31}$ cm and the radius of "outside" motion $\sim 10^{-10}$ cm.

The four structure of double motion is available for an electron. Let us schematically depict these structures as:

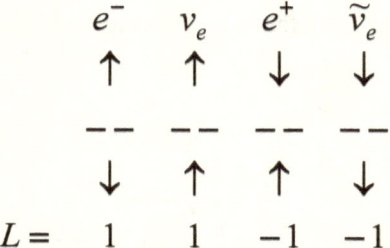

where a bottom arrow indicates the pointing of inside motion from center or to center and a top arrow indicates the pointing of outside motion. Imagine these structures as leptons of zero-level with corresponding values of lepton charge $L$.

For intermediate bosons $W^{\pm}$ and $Z_0$, ($n = 6$) this structure is:

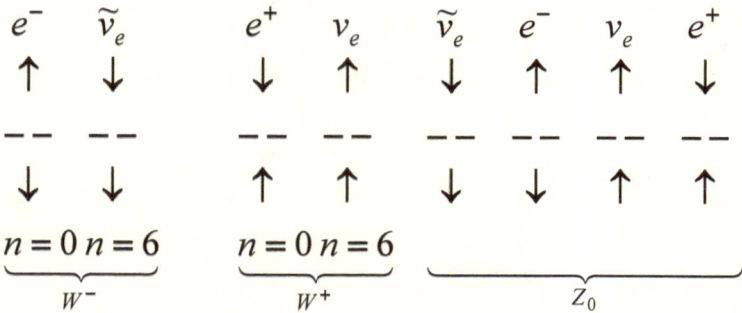

For component particles e.g. nucleons this structure is

$$L = 1, \quad \overline{L} = 2 \quad \prec$$

where $u$ and $d$ quarks can be concieved as of

$$\begin{cases} u(0,1,1) \\ d(1,0,1) \end{cases}$$

$n_1 \; n_2 \; n_4$

and value $n_0$ is responsible of the "color" of quark. Quark $d$ has color $\tilde{v}_e, e^+, e^-$ and quark $u^-$ has color $\tilde{v}_e, e^+, v_e$.

For the nucleons and antinucleons, the four combinations from four values by three values are may be made. Color of $e^-$ and $v_e$ has identical lepton number $L = 1$ but different charge stations.

The decay takes play by scheme: $n_0 \rightarrow p^+ + W^- \rightarrow p^+ + e^- + \tilde{v}_e$

1) $e^- \rightarrow W^- + v_e$.

2) $W^- \rightarrow e^- + \tilde{v}$

(The decay of $d$- quark of color «$e^-$» in to $u$-quark of color «$v_e$» and $W^-$ - boson that is $d \rightarrow u + W^-$).

$$e^- \qquad\qquad v_e \quad \tilde{v}_e \quad e^-$$

The masses of proton and neutron can be expressed as

$$m_p = \left[ \overline{L}\left(\overline{L} + L\right)\cdot n_1 \cdot 1{,}4 + L\left(\overline{L} + L\right)\left(\overline{L} - L\right)\cdot n_2 \cdot 3.8 \right]\cdot 207 m_e,$$

$$m_n = \left[ L\left(\overline{L} + L\right)\cdot n_1 \cdot 1{,}4 + \overline{L}\left(\overline{L} + L\right)\left(\overline{L} - L\right)\cdot n_2 \cdot 3.8 \right]\cdot 207 m_e,$$

where $\overline{L} = 2$ and $L = 1$ for nucleon, $n_1 = 1$, $n_2 = 2$ for proton and $n_1 = 2$, $n_2 = 1$ for neutron.

Thus, we obtain that

$$m_p = \left[ \overline{L}\cdot 3 \cdot n_1 \cdot 1.4 + L\cdot 3 \cdot n_2 \cdot 3.8 \right]\cdot 207 = 6(1.4 + 3.8)207 = 31.2 \cdot 207,$$

$$m_n = \left[ L\cdot 3 \cdot n_1 \cdot 1.4 + \overline{L}\cdot 3 \cdot n_2 \cdot 3.8 \right]\cdot 207 = 6(1.4 + 3.8)207 = 31.2 \cdot 207,$$

that is more than three times larger than mass of nucleon. Hence, more than 2/3 of this energy is bond energy of quarks in nucleon. Thus the mass of nucleon is equal to 1/3 of this energy minus the bond energy of single pion ($\sim$ 150 MeV).

For the $\pi$ - mesons the structure is:

$$\begin{cases} \overline{L} = 1 \\ L = 1 \end{cases}$$

$$u(0,1,1)$$
$$d(1,0,1)$$

72

The mass of $\pi^+$ and $\pi^-$ meson is:

$$m_{\pi^+} = m_{\pi^-} = (1 \cdot 2 \cdot n_1 \cdot 1.4 + 0 \cdot 3 \cdot n_2 \cdot 3.8) \cdot 207 m_e ,$$

that is two times larger than mass of meson. Thus, 1/2 of this energy is the bond energy of quarks in meson.

The difference of $\left| \overline{L} - L \right|$ may be equal to 1, for nucleon or 0 for meson.

The decay of pion takes play by scheme:

$$\pi^- \rightarrow \mu^- + \widetilde{v}_\mu , \ \mu(u), \ \widetilde{v}(d) \text{ see below.}$$

The muon and muon neutrino have the following quarkless "structure" :

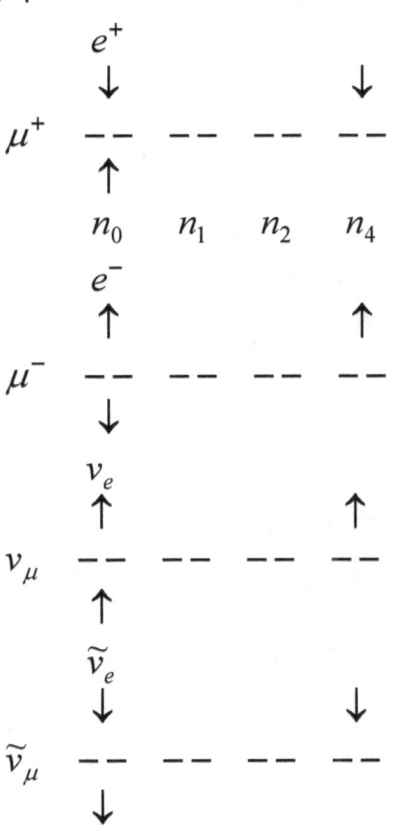

73

The decay of muon takes play by scheme:

$$\mu^- \rightarrow v_\mu + W^- \rightarrow v_\mu + e^- + \tilde{v}_e .$$

The electron neutrino gains the rest mass and escapes a particle at the instant of transitions $n_2 \leftrightarrow n_1$ (see scheme for nucleons). The evaluation of rest mass for neutrino as the variation of gravitational energy of virtual particle $m_{vir} = 1.3 \cdot 10^{19} \, m_e$ ($n = 2$) in the gravitational field of elec-

tron $\dfrac{Gm_e m_{vir\,(n=2)}}{r_2} \cong m_{0v_e}$ , where $r_2 \sim 2.5 \cdot 10^{-31}$ cm, will be in the

region of 1.8 eV, for $m_{0v_e}$ .

Not counting corrections introducing by $m_{vir\,4,5}$ ($n = 4,5$), that is

$$\frac{m_{0v_\mu} - m_{0v_e}}{m_{0v_e}} \sim 10^{-8} , \quad \frac{m_{0v_\tau} - m_{0v_e}}{m_{0v_e}} \sim 10^{-12} ,$$

the rest mass $v_\mu$ and $v_\tau$ is close equal to $m_{0v_e}$ .

## 5. The cosmological constant (the density of vacuum) and the relationship between the cosmic energies

1. Recent astronomical reseach has shown that the density of vacuum is

$$\rho_V \cong 2 \cdot 10^{-123} \, \rho_{pl} ,$$

where $\rho_{pl}$ is Planck density.

From the preceding, the density of <u>baryon</u> substance is would be expressible as:

$$\rho_B = \frac{M_U}{V_U} = \frac{\alpha \cdot m_{\text{pl}}}{\left(\alpha \cdot \dfrac{l_{\text{pl}}}{2}\right)^3} = \frac{8}{\alpha^2 \cdot \rho_{\text{pl}}},$$

where $\alpha = e^{144} \cong 3.5 \cdot 10^{62}$ that is $\rho_B = \dfrac{8}{10^{125}} \rho_{\text{pl}}$ and it is smaller than the above- mentioned density of vacuum by a factor of 25.

However, notice that the $\rho_V$ is variable value. It is invariable up to $\sim$ $10^{31}$ cm and then it will fall as $\dfrac{1}{R^3}$.

2. According to this recent astronomical research, the ratio of density of vacuum to critical density, that is its contribution to the total mass, is

$$\Omega_V = \frac{\rho_V}{\rho_c} = 0.7 \pm 0.1,$$

the contribution of dark substance of galaxies is

$$\Omega_D = \frac{\rho_D}{\rho_c} = 0,3 \pm 0,1,$$

and contribution of baryons is

$$\Omega_B = \frac{\rho_B}{\rho_c} = 0.03 \pm 0.01$$

Then, the contribution of fourth component of cosmic medium that is radiation or ultra relativistic medium is equal to

$$\Omega_R = \frac{\rho_R}{\rho_c} = 0.8 \cdot 10^{-5} k, \qquad 1 < k < 10 \div 30.$$

These ratios may be explained in terms of the above mentioned hypothesis.

3. Let us turn back to Table 1. Suppose that, at $n = 12$, the particles with radius $\sim 10^{31}$ cm produce a repulsion field (space curvature is negative). This is the so-called vacuum. At $n = 11$, the particles with radius $\sim 10^{21}$ cm produce an attraction field (space curvature is positive). This is the so-called dark substance of galaxies.

At $n = 10$, the particles with radius $\sim 10^{13}$ cm produce a repulsion field too (space curvature is negative). These fields are responsible for the origination of planetary systems (the before planet cloud).

At $n < 10$, this alternation is being continued. It is conceivable that theseparticles (fields) are responsible for the formation of protoplanetary bodies with the asteroid dimensions from the before planet cloud ( at $n = 9$), for the nanoeffects (Kazimir's forces) (at $n = 8$) and for the synthesis of heavy elements (at $n = 7$).

Let us make round estimates of amount and total mass of these particles ($n = 12, 11, 10$) which are originated at the instant the "Big Bang" takes place:

$$N_{12} \sim \frac{T_{12}}{\tau_{12}} \cdot \beta \cdot N \,,$$

where $T_{12}$ is time $t_0$ until the completion of this stage (era);

$$T_{12} \sim 10^{-6}\, \text{s}, \qquad \tau_{12} \sim 33.1 \cdot \tau_{pl} = 1.7 \cdot 10^{-42}\, \text{s},$$

$\tau_{12}$ – characteristic time of process (the birth of particles),

$\beta = 1.3 \cdot 10^9$ is factor of baryon asymmetry since the process takes place before the annihilation of baryons,

$N = 4.4 \cdot 10^{81}$ is amount of baryons (quarks) in processes quark $\leftrightarrow$ lepton,

$$N_{12} \sim \frac{10^{-6}\, s}{1.7 \cdot 10^{-42}\, s} \cdot 1.3 \cdot 10^9 \cdot 4.4 \cdot 10^{81} \cong 3 \cdot 10^{126} \,.$$

In similar manner, at $n = 11$ (the transitions quark $\leftrightarrow$ baryon, hadron era)

$$N_{11} \sim \frac{1.5 \cdot 10^{-4}\, s}{1.7 \cdot 10^{-29}\, s} \cdot 1.3 \cdot 10^9 \cdot 4.4 \cdot 10^{81} \cong 10^{116} \,.$$

At $n = 10$ (transitions $n \leftrightarrow p$, lepton era)

$$N_{10} \sim \frac{1s}{10^{-20}\, s} \cdot 4.4 \cdot 10^{81} \cong 4 \cdot 10^{101} \,,$$

where $10^{-20} s = \tau_{ew}$ is time of electroweak processes and when the factor $\beta$ is eliminated.

The ratio between masses of these particles is

$$\frac{m_{12}}{m_{11}} \sim \frac{2 \cdot 10^{-42}\, m_e}{2 \cdot 10^{-32}\, m_e} = 10^{-10}, \frac{m_{11}}{m_{10}} \sim \frac{2 \cdot 10^{-32}\, m_e}{2.6 \cdot 10^{-23}\, m_e} = 10^{-9},$$

and then the ratio between the total contributions is

$$\frac{M_{12}}{M_{11}} \sim 2 \div 4, \frac{M_{10}}{M_{11}} \sim 2 \cdot 10^{-5}, M = m \cdot N.$$

4. Let us turn back to a consideration of baryons. Baryon is not simple fivedimensional sphere of size ~ $10^{-31}$ cm. It is particle of substance having the fields of strong, Coulomb and weak interactions. For fivedimensional sphere (the space inside is sixdimensional and has 3 +1+ mass + spin) the surface area is

$$\pi^3 R^5 = S_5 = 31.006 R^5,$$

that is the compactification or reduction factor is equal to 31. The vacuum surrounding baryons is fivedimensional (3+1+ mass or curvature).

Thus, it is felt that after the decay substance the space curvature will be fully compensated as initially the total curvature is equal to zero. That is at decay (i.e. reducing the baryon dimensionality by 1) the baryons are to be annihilated with space. The mass of these baryons are $\dfrac{1}{31}$ part (or 3.2%) of total mass of the Universe. The disturbance of bilateral symmetry in the decay of matter particles is a manifestation of this effect. To ensure the compensation of general curvature the attraction and repulsion fields in baryons are to be in energetic and integral balance with the corresponding fields of the Universe.

The Universe is electrically neutral i.e. .electric charge is fully compensated. The baryon charge is not compensated.

Let us turn back to Table 1. Suppose that the quark $\begin{pmatrix} d \\ 1.4 \end{pmatrix}$ creates the attraction field at $n =1$ and the quark $\begin{pmatrix} u \\ 3.8 \end{pmatrix}$ creates the repulsion field at $n = 2$. At $n = 3$ quark $\begin{pmatrix} c \\ 20 \end{pmatrix}$ creates the attraction field too. Thus, the $u$ and $d$ quarks are liable to form combination wherein the size of attraction area $\left( \sim \dfrac{1}{m_d} \sim 1.9 \cdot 10^{-13} \text{ cm} \right)$ stands out above the size of repulsion area $\left( \sim \dfrac{1}{m_u} \sim 0.7 \cdot 10^{-13} \text{ cm} \right)$ and the repulsive potential

i.e. core is large than attractive potential by a factor $\dfrac{W(-)}{W(+)} = \dfrac{3.8}{1.4} = e$.

As total curvature of the Universe is equal to zero and space (space energies) is "antimatter" for baryons, the given below ratio may be written:

$$\frac{\rho_V}{\rho_D} = \frac{W(-)}{W(+)} = e.$$

The constant of weak interaction is five orders less than the constant of strong interaction, this is in accordance with the ratio:

$$\frac{\rho_R}{\rho_V} \ll 1.$$

Thus, doing not take into account the contribution of $\rho_R$, it may be roughly written as:

$$\begin{cases} \Omega_V + \Omega_D + \Omega_B \cong 1, \\ \dfrac{\Omega_V}{\Omega_D} = e, \end{cases}$$

$$\Omega_R \ll 1, \quad \Omega_B = \frac{1}{\pi^3} \cong \frac{1}{31}.$$

We can apply this relationships to give

$$\begin{cases} \Omega_V = \dfrac{e\left(1 - \dfrac{1}{\pi^3}\right)}{1 + e} = 0.7075 \cong 70.8\%, \\ \Omega_D \cong 0.26 = 26\%, \\ \Omega_B \cong 3.2\%, \end{cases}$$

this is in good agreement with the last experimental data.

## Conclusion

An effort is made to associate the relationships controlling the world of elementary particles and the Universe as a whole. The standard model of elementary particles adequately describes observations. However, this model postulates the values of particle masses and does not use the concepts explaining relationship among these masses. Hence, it is significantly describing theory and its development is previously advanced by experimental results. It is suggested that this model will be faced problems of explanation of facts which will be occurred in the course of the energy increasing of collaiders. There are failed detections of Higgs' bosons and new particles in range about 6 TeV.

It is also unlikely that progress in search for exotic massive particles forming the dark matter in surrounding space will be made as this matter is surrounding space with some energy density.

This work is not of course fundamental theory for all properties of elementary particles and the Universe. The hypothesis supporting the body of known facts is only reported. It is conceivable that a some integration of the standard model of elementary particles with modern cosmology and hypothesis outlined above will furnish insights into the nature of problem.

Khmelnik S. I., Khmelnik M. I.

# Analysis of Energy Processes in Searle's Generator

## Annotation

The Searle's generator is treated from the point of view of energy transformation processes. On the base of experimental data and heat exchange theory, it is shown that the interaction of constant magnets with the environment may serve as an energy source. Based on this fact a detailed analysis of energy process is performed in this article, with construction of an appropriate differential equation. The methods of its solution are considered, as well as the results of working calculations of the known experimental device.

## Contents

## 1. Introduction

The Searle's generator is known from internet publications [1-3]. As a first approximation, the Searle's generator can be envisioned as a structural unit reminding a ball-bearing – metal cylinders rolling around a metal rim. Its design is based on the discovery of *Searle's technological effect,* the essence of which is that the magnetizations of a certain material by direct current "with a touch" of high-frequency component creates a multitude of magnetic poles on the surface of this material. The rim-stator is magnetized in such a way that on each of its boundary circles there is a multitude of one-sign magnetic poles. The rotor's cylinders are also magnetized in such way that there is a multitude of magnetic poles on their circle. *The Searle's kinematic effect* lies in the fact, that after the rotor's acceleration with the aid of external motor and after the latter's switching-off, the rotor continues to accelerate to a high speed.

First experiments with these generators were conducted already in 1946. The experiments with Searle's generator and with similar devices of Roshchin-Godin are described in detail in [1-6]. However, the source of energy and the sources of driving forces in the generator haven't been discovered till today.

So, in the analysis of the experiments with the Searle's generator, [1-6] two main questions emerge: 1) What is the origin of the driving force and 2) where the generator takes the energy from? In [7] we have shown that the driving force cannot be a result only of force interaction between the constant magnets of the rotor, and so the first question remains an open issue. Here we are analyzing the second question and give an answer to it. The results of works [8, 9] are being generalized and further developed.

Before considering the Searle's generator we must call attention to some other facts which rank with it from the point of view of the raised question.

It may be assumed that constant magnet in certain conditions can be a transformer of internal energy of the environment into other forms of energy. It becomes most evident in light of the existence of magnetocaloric effect – MCE [10]. MCE is the capacity of any magnetic material to change its temperature under of a magnetic field impact. The maximal magnitude of MCE is reached in ferromagnetic materials. Some materials (for instance, gadolinium) increase their temperature quite substantially. At present this effect is being used even for home refrigerators manufacturing. The estimates show [10], that the use of "magnetic" fridges allows to reduce the USA energy consumption by 5%.

According to the law of energy conservation we should recognize that

1) or the constant magnet is a certain catalyst, aiding to ferromagnetic material (for instance, to gadolinium) to accumulate the internal energy of the environment;

2) or that the magnetic energy of a constant magnet transforms into internal energy of ferromagnetic materials; in this case we must also assume that

| | |
|---|---|
| The energy of constant magnet is replenished at the expense of the internal energy of the environment | (A) |

In any case

| | |
|---|---|
| The energy of environment is being transformed (in addition to heat exchange) by the constant magnet into the internal energy of ferromagnetic materials | (B) |

## 2. The Energy Transformation in the Searle's Generator

By analogy with the above said it may be assumed that in Searle's generator

| |
|---|
| The internal energy of environment is being transformed by the constant magnets to the kinetic energy of the rotor,            (C) |

This means that the Searle generator is a transformer of the internal energy of the environment into kinetic energy. It may be confirmed by the fact that in the experiments of Searle, Roshchin-Godin [1, 2, 6] temperature decrease by 6-8 grades has been marked. Further we shall refer to the experiment with Searle's generator described in [6], to substantiate one or other assumptions, because there we found the fullest data.

From the assumption (C) it follows that in the analysis of Searle's generator the following forms of energy should be considered:

1. magnetic energy of constant magnets $W$ ,
2. internal energy of constant magnets $U$ ,
3. kinetic energy of rotor $K$ ,
4. energy consumption of starting motor and the generator's loading $A$ ,
5. internal energy of the environment; its variation will be denoted as $Q$ .

The following energy transformations take place here

$Q \Rightarrow \Delta U$ : heat transmission from the environment to Searle's generator, which is confirmed by the existence of the observed temperature difference;

$\Delta U \Rightarrow \Delta W$ : this transformation we assume to exist, without proposing any specific mechanism of such transformation; the further described experiments will substantiate it; evidently the constant magnets, when cooling, have had lost some part of their internal energy as a result of processes taking place in the crystal lattice of the constant magnet materials; this is precisely the energy that was transformed into magnetic energy of constant magnets.

$\Delta W \Rightarrow \Delta K$ , $\Delta W \Rightarrow A$ : we should accept the existence of these transformations simply for the reason that the rotor of Searle's generator rotates, and <u>constant magnets</u> are an indispensable component of this generator.

Thus, there exists the following succession of energy transformations: $Q \Rightarrow \Delta U \Rightarrow \Delta W \Rightarrow K + A$. To simplify the quantitative analysis we shall exclude the intermediate transformations of magnetic energy in constant magnets, and so this succession will take the form: $Q \Rightarrow \Delta U \Rightarrow K + A$.

## 2. Differential Equation of Energy Process

Based on these energy transformations we shall further deduce a differential equation describing the energy processes in the generator. This equation permits to analyze the generator's behavior in various circumstances, and to perform some "virtual" experiments.

For a small time period $dt$ the energy conservation law may be written as the following equation:

$$dA_p + dU + dQ = dK + dA_n, \tag{1}$$

where

$dA_p$ - the starting motor energy variation,

$dA_n$ - the load energy variation,

$dU$ - the constant magnets internal energy variation,

$dK$ – the rotor kinetic energy variation,

$dQ$ - the variation of energy received in the process of heat exchange with the environment

Further we have:

$$dA_p = Gdt, \tag{2}$$

$$dU = c \cdot m \cdot dT, \tag{3}$$

where

$m$ - the mass of generator's constant magnets,

$c$ - the specific heat of constant magnets material,

$T$ – the absolute temperature of generator's constant magnets,

$G$ – the power of **motor speeding up the rotor**.

From the Fourier's formula for convective heat exchange [11] with the environment we have

$$dQ = \alpha \cdot S \cdot (T_o - T)dt, \tag{4}$$

where

$S$ – the constant magnets surface,

$T_o$ – the absolute temperature of the environment,

$\alpha$ - the heat emission coefficient, depending on the environment characteristics and on the process of motion around the generator (for instance, on the room's ventilation).

The energy **contributed by the rotor to the load** is equal to

$$dA_n = Pdt .$$
(5)

The kinetic energy variation is equal to

$$dK = d\left(0.5 m_r v^2\right) = m_r \cdot v \cdot dv .$$
(6)

Here

$P$ – the load power of the rotor shaft,

$m_r$ – the rotor mass,

$v$ – the rotor rollers linear velocity.

Then

$$dK = m_r \cdot v \cdot dv .$$
(7)

From (1-6) we get the following differential equation of the process

$$\left\{ \begin{array}{l} G(t)dt + cm\dfrac{dT(v)}{dv}dv + \alpha S(T_o - T(v))dt \\ - m_r \cdot v \cdot dv - P(t)dt \end{array} \right\} = 0 .$$
(8)

All intermediate transformation involving magnetic energy was excluded from this equation. However we must take into account, that the generator's energy cannot exceed the magnetic energy of the rotor's and stator's constant magnets, which may be computed by the formula

$$W_m = \frac{V_m BH_m}{2} .$$

The intensity $H_m$ in the body of constant magnet with rectangular hysteresis loop may vary in a wide range for a constant induction $B$. Consequently, the constant magnets energy may vary in a wide range for a constant induction $B$.

It is convenient to study the equation (8) when written in non-dimensional variables. To do this we shall introduce certain etalon time $t_o$ and velocity $v_o$:

$$v_o = \sqrt{cT_o} ,$$
(9)

$$t_o = R/v_o ,$$
(10)

where $R$ is the rotor's radius. Then we shall introduce non-dimensional time and velocity

$$\tau = t/t_o ,$$  (11.1)

$$\xi = v/v_o$$  (11.2)

or

$$\xi = \pi \cdot n \cdot R \big/ 30 v_o ,$$  (11.3)

where $n$ – revolutions per minute.

Each item in (8) has the dimension of energy. Therefore we shall introduce etalon energy

$$Q_o = mcT_o$$  (12)

and divide by it all the terms of the equation (8). Then the equation (8) will take the form

$$\left\{ \begin{array}{l} \overline{G}(\tau)d\tau + \dfrac{dk_1(\xi)}{d\xi}d\xi + k_2\big(1 - \overline{T}(\xi)\big)d\tau \\[2mm] - k_4\xi \cdot d\xi - \overline{P}(\tau)d\tau \end{array} \right\} = 0 ,$$  (16)

where

$$\overline{T}(\xi) = \frac{T(v)}{T_o}, \quad k_2 = \frac{\alpha S t_o}{mc}, \quad \overline{G}(\tau) = \frac{t_o G(t)}{Q_o},$$

$$k_4 = \frac{m_r v_o^2}{Q_o}, \quad \overline{P}(\tau) = \frac{t_o P(t)}{Q_o}.$$  (17)

The non-dimensional constants $k_2$ and $k_4$ depend only on the constant parameters of generator and environment. They may be called "critical numbers" of this process, by analogy with critical numbers in other spheres (for example, Reynolds number in hydrodynamics, Mac number in aerodynamics etc.). The number $k_2$ describes heat exchange between the constant magnets and the environment, and the number $k_4$ characterizes the rotor's inertia.

Equation (16) allows to determine the dependence $\xi(\tau)$. After that we will be able to find the dependence of $v(t)$ by (11).

Let us denote

$$\overline{T}'(\xi) = \frac{d\overline{T}(\xi)}{d\xi}$$

and from (16) we shall find

$$\frac{d\xi}{d\tau} = H(\xi,\tau) = \left\{ \frac{\overline{G}(\tau) + k_2(1 - \overline{T}(\xi))d\tau - \overline{P}(\tau)}{k_4\xi \cdot d\xi - \overline{T}'(\xi)} \right\}. \quad (19)$$

## 3. About the Choice of Dependence $\overline{T}(\xi)$

In the equation (19) the speed and the time are unknown. Therefore it should be supplemented by another equation. For this purpose we shall consider a function $\overline{T}(\xi)$, corresponding to the dependence $T(v)$. The form of this function is not known. We know only the minimal value $T_{min}$ of $T$, i.e. maximal value of $\Delta T = T_0 - T_{min}$. This value of $\Delta T$ corresponds to minimal value of $\overline{T}(\xi)$, which will be denoted as

$$\overline{T}_{min} = 1 - \delta,$$

where

$$\delta = \Delta T / T_o, \quad (20)$$

as $\overline{T}_{min} = \dfrac{T_{min}}{T_o} = \dfrac{T_o - \Delta T}{T_o} = 1 - \delta.$

We shall choose the function $\overline{T}(\xi)$ so as to satisfy the following conditions

1) $\overline{T} = 1 - \delta$ for $\xi = \infty$,

2) $\overline{T}(0) = 1$ and $\overline{T}(\xi)$ is monotone decreasing.

Since the number of such functions in unlimited, we shall choose only most simple dependences $\overline{T}(\xi)$ and among them – those that agree better with the experimental data. We shall take the following function in this capacity

$$\overline{T}(\xi) = 1 - \delta + \delta / (\delta\gamma\xi + 1)^{\beta}. \quad (21)$$

Wherefrom we can find:

$$\overline{T}'(\xi) = -\left(\gamma\beta\delta^2\right) / (1 + \gamma\delta\xi)^{\beta+1}. \quad (22)$$

The constants $\beta$, $\gamma$ will be chosen by comparing the calculations results with experimental data.

Taking also into account (20, 10, 11, 17) and

$$n = \frac{30v}{\pi R} \quad (22.1)$$

we shall find

$$T(n) = T_o\left(1 - \delta + \delta/(\delta\gamma\xi + 1)^\beta\right).$$                    (22.2)

Figure 0 shows the dependence (22.2) for various values of the constants $\gamma$, $\beta$.

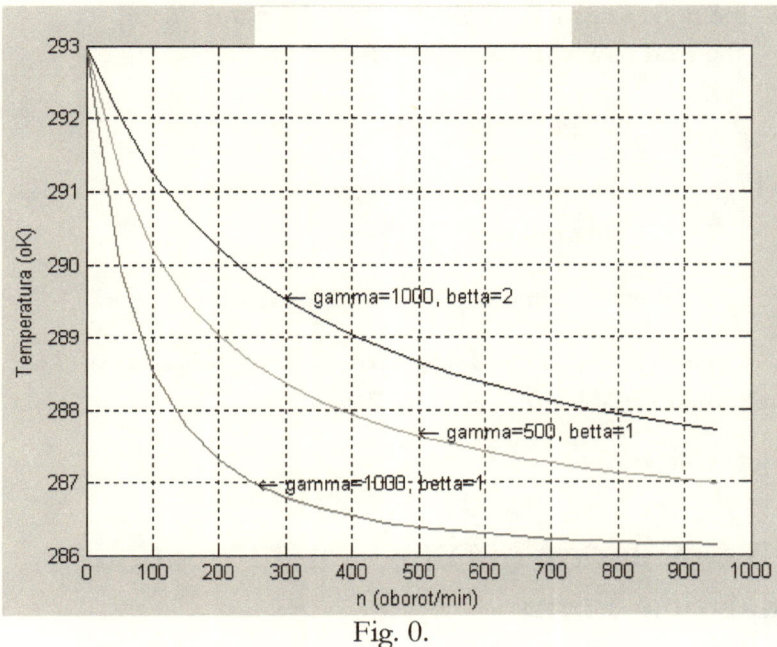

Fig. 0.

In particular, for $\beta = 1$ from (22.2) we have:

$$\left(\frac{T}{T_o} - 1 + \delta\right) = \delta/(\delta\gamma\xi + 1)$$

or, taking into account (9, 11.3),

$$n = \left(-\frac{1}{\delta} + 1\middle/\left(\frac{T}{T_o} - 1 + \delta\right)\right)\frac{30v_o}{\pi R\gamma},$$                    (22.4)

$$T = T_o\left(1 - \delta + \delta\middle/\left(1 + \frac{\delta\gamma\pi R}{30\sqrt{cT_o}} \cdot n\right)^\beta\right),$$                    (22.5)

Let us now consider the power balance equation. From (8) we find:

$$G + cm\frac{dT}{dt} + \alpha S(T_o - T) - m_r \cdot v\frac{dv}{dt} - P = 0,$$                    (22.6)

or, taking into account (22.1),

$$G + cm\frac{dT}{dt} + \alpha S(T_o - T) - \frac{m_r \cdot \pi^2 R^2 n}{900} \cdot \frac{dn}{dt} - P = 0 \text{,(22.7)}$$

where

$G$ – the power of **motor speeding up the rotor.**

$P$ – the load power of the rotor shaft,

$m_r \cdot v\dfrac{dv}{dt}$ – «kinetic» rotor power,

$\alpha S(T_o - T)$ - the power of convective heat exchange with the environment

$cm\dfrac{dT}{dt}$ - power of the magnets internal energy.

So, the generator dynamics is described by two equations (22.5, 22.6) with unknown variables $T$, $n$. The dependence of $T$ on $n$ -according to the given dependence of $\overline{T}$ on $\xi$.

## 4. The Solution of Differential Equation in Analytical Form

Let us now consider a particular (but fairly general) case, when the main equation (19) may be solved in analytical form. It will be the case when the external power $G = \text{const}$ till a certain moment $t_n$ and $G = 0$ for $t > t_m$, and the load power $P$ is constant with respect to time. Therefore, in this case $\overline{G}$, $\overline{P}$ are constants. Then the equation (19) may be written in the following form:

$$\frac{d\tau}{d\xi} = L(\overline{G},\xi) = \left\{ \frac{k_4\xi - \overline{T}'(\xi)}{\overline{G} + k_2(1 - \overline{T}(\xi)) - \overline{P}} \right\}. \qquad (23)$$

In this case the dependence $\tau(\xi)$ can be found by finite integration. All the process of motion may be divided in two stages

1) acceleration process under the influence of external force $(\overline{G} \neq 0)$,

2) process of spontaneous motion $(\overline{G} = 0)$.

On the first stage from (23) we have

$$\tau = \int_0^{\xi} L(\overline{G}, \xi) d\xi = \Phi(k_3, \xi), \quad \xi \leq \xi_m, \qquad (24)$$

where $\xi_m$ is the value of $\xi$ for $\tau = \tau_m$ (corresponding to $t = t_m$). From that we can find the dependence $\xi(\tau)$ in implicit, but analytical form. The value of $\xi_m$ may be found by solving numerically the equation

$$\Phi(k_3, \xi_m) = \tau_m, \qquad (25)$$

On the second stage $\tau > \tau_m$ the equation (19) will take the following form

$$\tau(\xi) = \tau_m + \int_{\xi_m}^{\xi} L(0, \xi) d\xi = \Phi(0, \xi). \qquad (26)$$

If for $\xi > \xi_m$ the equation (26) gives the values $\tau < \tau_m$, this means that the speed after acceleration decreases, and the calculations by (26) should be performed with $\xi < \xi_m$. The same conclusion may be reached in the case of $L(0, \xi) < 0$ for $\xi < \xi_m$.

## 5. The Solution of Differential Equation in Parametric Form

If $\overline{T}(\xi)$ is defined by (21), then from (23) we shall get:

$$\frac{d\tau}{d\xi} = \frac{k_4 \xi (1 + \delta\gamma\xi)^{\beta+1} + \gamma\beta\delta^2}{(\overline{G} - \overline{P} + k_2\delta)(1 + \delta\gamma\xi)^{\beta+1} - k_2\delta(1 + \delta\gamma\xi)}. \qquad (27)$$

and $\dfrac{d\tau}{d\xi}$ is equal to a fraction, whose numerator and denominator are polynomials of the variable $\xi$. Now it will be more convenient to turn to the variable

$$u = (1 + \delta\gamma\xi). \qquad (28)$$

then we shall have

$$\frac{d\tau}{d\xi} = A(\overline{G}) \frac{(u-1)u^{\beta+1} + B}{u^{\beta+1} - D(\overline{G})u} = H(\overline{G}, u), \qquad (29)$$

where

$$A(\overline{G}) = \frac{k_4}{\gamma^2 \delta^2 (\overline{G} - \overline{P} + k_2 \delta)},$$

$$B = \frac{\gamma^2 \delta^3 \beta}{k_4},$$

$$D(\overline{G}) = \frac{k_2 \delta}{(\overline{G} - \overline{P} + k_2 \delta)}.$$

The formulas (28) and (29) give the dependence $\dfrac{d\tau}{d\xi}$ on $\xi$ in parametric form. Then the dependence $\xi$ on $\tau$, defined by the formulas (24) and (26) may be written in parametric form as follows:

in the speeding section

$$\tau(u) = \int_1^u H(\overline{G}, u) du = \Psi(\overline{G}, u), \tag{30}$$

in the section of spontaneous motion

$$\tau(u) = \tau_m + \int_{u_m}^u H(0, u) du. \tag{31}$$

These formulas together with (28) define the dependence $\tau(\xi)$. The value $u_m$ may be found similarly to (25) from the equation

$$\tau_m = \Psi(\overline{G}, u_m). \tag{32}$$

Let us consider now a particular case of $\beta = 1$. Then

$$\frac{d\tau}{d\xi} = H(\overline{G}, u) = A(\overline{G}) \frac{u^3 - u^2 + B}{u(u - D(\overline{G}))} \tag{33}$$

and after integrating we shall get

in the speeding section

$$\tau = A(\overline{G}) \left[ \begin{array}{l} \dfrac{u^2 - 1}{2} + (D-1)(u-1) - \dfrac{B}{D}\ln(u) \\[2mm] + \left(D^2 - D + \dfrac{B}{D}\right)\ln\dfrac{u-D}{1-D} \end{array} \right], \tag{34}$$

in the section of spontaneous motion

$$
\tau = \tau_m + A(0)\left[\begin{array}{l}\dfrac{u^2 - u_m^2}{2} + (D-1)(u - u_m) - \dfrac{B}{D}\ln\dfrac{u}{u_m} \\ + \left(D^2 - D + \dfrac{B}{D}\right)\ln\dfrac{u - D}{u_m - D}\end{array}\right]. \tag{35}
$$

## 6. System without Magnets

Of some interest is also the comparison between the performance of Searle's generator on the speeding section with the same device's performance with demagnetized magnets. For such comparison we must calculate with the aid of the above cited formulas the speeding time $t_m$ for Searle's generator and the similar values of $t_{om}$ for the compared device with the same number of rotations per minute after speeding $n$.

Hence due to the absence of cooling under the influence of magnetic field $\overline{T} = 1$, $\overline{T}' = 0$ and the equation (23) will take the form

$$
\frac{d\tau}{d\xi} = \frac{k_4 \xi}{\overline{G} - \overline{P}}. \tag{36}
$$

From this, after integration, we get

In the speeding section $\left(\overline{G} \neq 0\right)$

$$
\xi = \sqrt{\frac{2\tau\left(\overline{G} - \overline{P}\right)}{k_4}}, \tag{37}
$$

In the section of spontaneous motion $\left(\overline{G} = 0\right)$

$$
\xi = \sqrt{\xi_m^2 - \frac{2\overline{P}}{k_4}(\tau - \tau_m)}. \tag{38}
$$

so we see that in the section of speeding the rotor's motion will be accelerated (but with decreasing acceleration), and in the section of spontaneous motion it will be slowing down. Notice that for constant speeding power $G$ the force exerting on the rotor will be decreasing with increasing speed, as $G = Fv$.

## 7. Steady State Mode

Let us consider the steady state mode, when speed and temperature do not change, and the speeding-up motor is switched off. In this case the load power is equal to generator power. From (22.6) we have

$$P = \alpha S(T_o - T).  \tag{41}$$

Using the formulas (22.5, 41) we shall be able to derive the dependence of temperature $T$ and load power $P$ on the number of rotations per minute $n$.

The above used method of calculation of self-excitation mode dynamics is suited for the case when the load power is constant. It is shown that in this case the speed increases infinitely. But actually the load power increases with rotation speed enhancement. It means that at a certain moment the steady state mode sets in. This is the mode for which the formulas (22.5, 41) may be used. Let us consider now the transition to steady state mode for variable load power, increasing (as was noted above) with the rotation speed growth. For this the equation system (22.5, 22.7) may be used, with power $P$ given as a dependence on the number of rotations per minute $n$. The equation (27), may also be used where $\xi$ is defined according to (11.3), and

$$\frac{d\tau}{d\xi} = \frac{30v_o}{\pi R t_o} \cdot \frac{dt}{dn},  \tag{42}$$

which follows from (11.1, 11.3). In this case the power $P$ may also be defined as a dependence on the number of rotations per minute $n$. Solving the equations system (27, 11.3, 42) will permit to build a graph of dependence of $\dfrac{dn}{dt}$ rotations number $n$ on time $t$ and dependence of $\dfrac{dn}{dt}$ on rotation number $n$ for variable power.

## 8. Experiments

In [3, 4, 5] the designs of Searle's generator and experiments with them are described. However in this description some constructive data necessary for our analysis are lacking. In [8] these experiments are being analyzed and completed by the necessary parameters. Further we shall use the data from [8], namely:

$$G = 7000\text{вт}, \quad m = 225\text{кг}, \quad \Delta T = 7\text{К}, \quad m_r = 4m,$$

$$T_o = 293K = 20°C, \quad c = 125\,\text{Дж}/(\text{кг}\cdot\text{К}),$$

$$\rho = 8000\,\text{кг}/\text{м}^3, \quad R = 0.5\text{м}, \quad S = 1.44\text{м}^2,$$

$$\gamma = 1000, \quad n_{\max} = 550\,\text{об}/\text{мин}, \quad \alpha = 222.$$

For these parameters

$$\delta = 7/293, \quad v_o = 191\text{м}/\text{сек}, \quad t_o = 0.26\text{сек}.$$

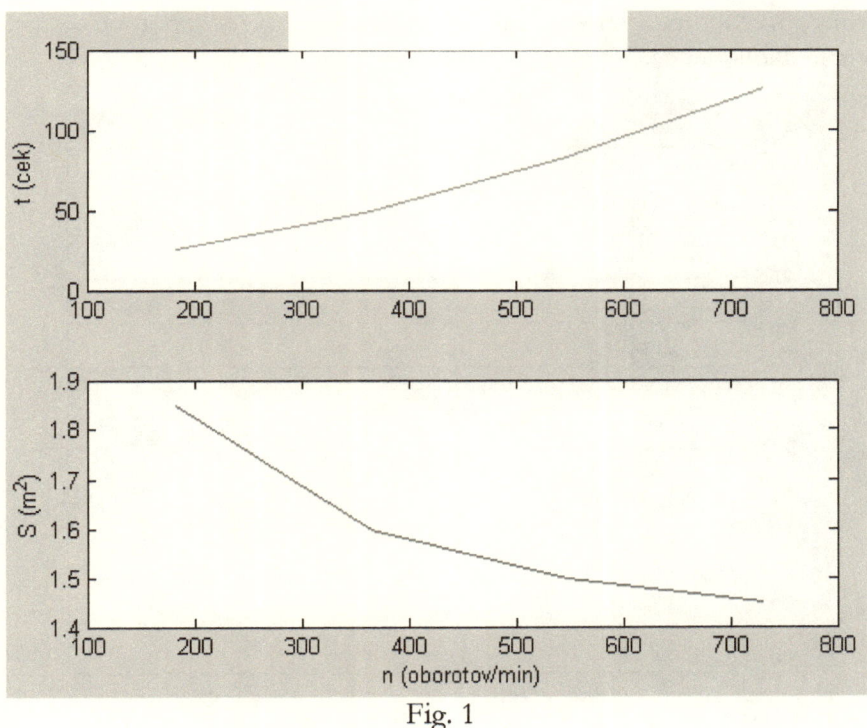

Fig. 1

Let us consider the modes of generator's operating

Acceleration mode

In this mode P=0, and the dependence of speed on time takes the form (34). Figure 1 shows the results of virtual experiments used to define the conditions of the generator's self-excitation. Figure 1 shows the values of the square $S$ and of the necessary minimal speeding time (at this moment the speeding motor is disconnected) for a given value of $S$, and also the number of rotations achieved by the generator. The calculations were done by formula (34), and the moment of speeding motor disconnection (i.e. setting $G = 0$) was determined by formula (29), when acceleration (for $G = 0$) changed its sign from «minus» to «plus». At this moment $\xi = \xi_{max}$. In this way it was discovered that for each value of magnets square $S$ there exist such minimal speeding velocity and rotation speed, that for their smaller values the self-excitation mode of a generator does not exist.

<u>Self-excitation Mode.</u>

In this mode $G = 0$ and the dependence of speed on time has the form (35). Figure 2 shows the dependence of speed on time and the inverse dependence.

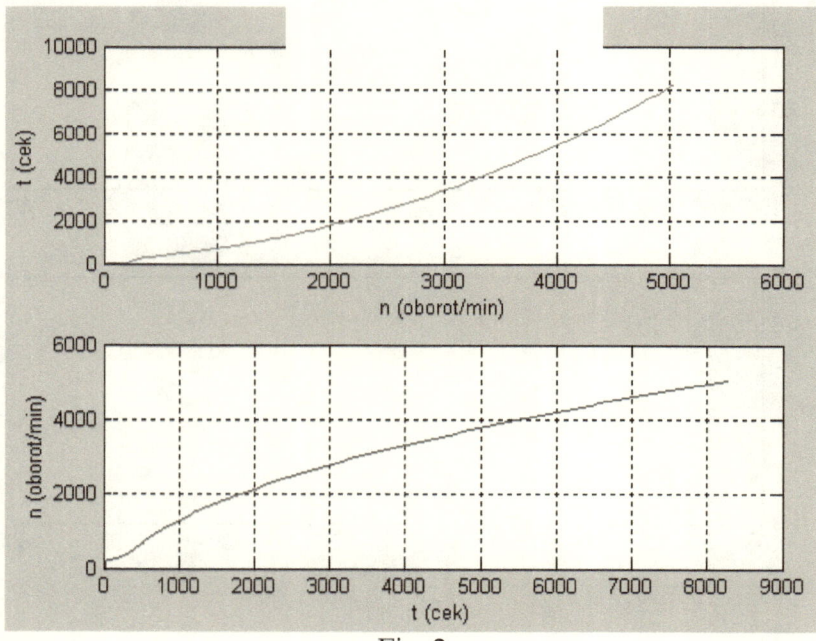

Fig. 2

<u>Steady State Mode.</u>

If it is known that the mode became steady state for a given rotations number $n$, then the temperature $T$ and the load power $P$ for this mode may be found from the equations (22.5, 22.7). Figure 3 shows the dependences of temperature $T$ and load power $P$ on the number of rotations per minute $n$ for the steady state mode.

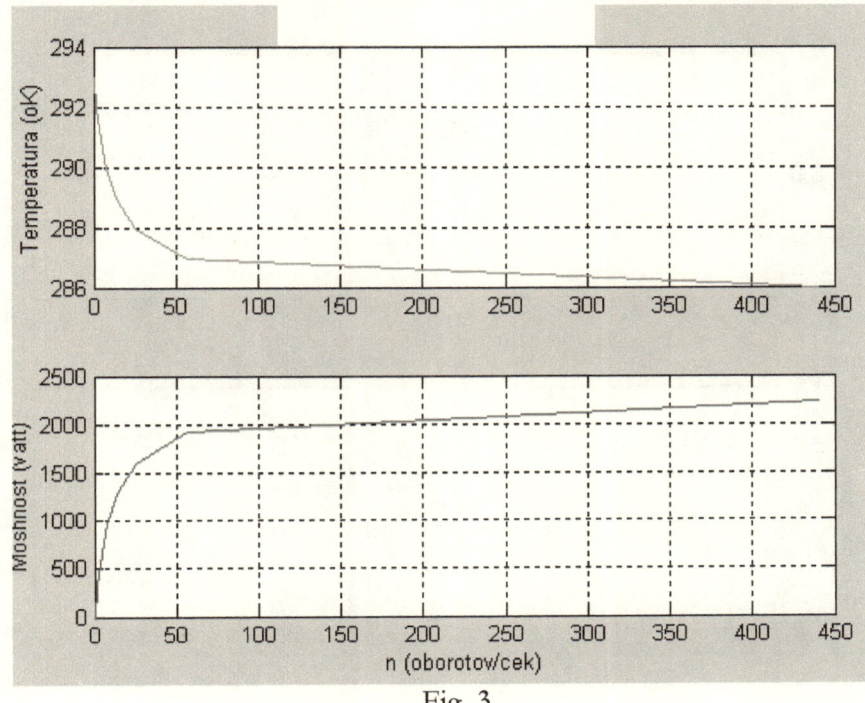

Fig. 3.

Let us assume now that the load power increases with the growth of rotations number. In the Fig. 4 an example of such dependence is shown4 in the lower right window. There the notation **GP** is used for the number **(G-P),** i.e. the case considered is such where the speeding motor changes from a motor mode to a power consumer mode. If it is known that the mode became steady state at the given rotation number per minute $n$, for this case then from the equations (22.5, 22.7) we may find the load power $P$ and acceleration $\dfrac{dn}{dt}$ as depending on rotation number and on time that passed from the moment of speeding motor start. Fig. 4 shows that the steady state mode starts at P=2000 and n=650, if the stated dependence of power on rotation number is actually valid.

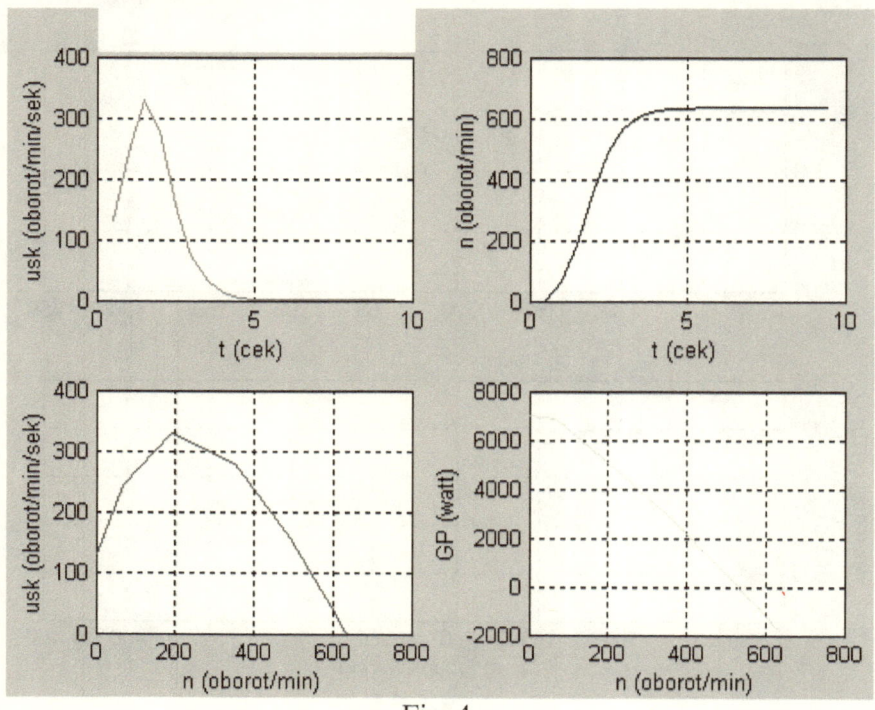

Fig. 4.

# References

1. Generator Based on Searle's Effect. SUSSEX University. Faculty of Engineering and Applied Sciences. Report SEG-002. (in Russian) http://ntpo.com/invention/invention2/23.shtml
2. S. Gunner Sendberg. Antigravitation. Searle's Effect. (in Russian)http://www.ufolog.nm.ru/artikles/searl.htm
3. Searl International Space Research Consortium http://www.sisrc.com/e.htm
4. John Searl Solution, http://www.searlsolution.com/investing.html
5. Roshchin V. V., Godin S. M. A Device for Mechanical Energy Production and a Method of Mechanical Energy Production, Patent of RF, H02N11/00, F03H5/00, 2000. (in Russian) http://macmep.h12.ru/roshin.htm
6. Roshchin V. V., Godin S. M. Experimental Investigation of Physical Effects in Dynamic Magnetic System. Letters to Journal of Theoretical Physics, 2000, volume 26, iss. 24_(in Russian). http://www.ioffe.rssi.ru/journals/pjtf/2000/24/p70-75.pdf

7. Khmelnik S. I. And Khmelnik M. I. To the Question of the Driving Force Source in the Searle's Generator. "Independent Authors Reports", publ. «DNA», printed in USA, Lulu Inc., ID 322884. Russia-Israel, 2006, issue 4 (in Russian).

8. Khmelnik S.I. and Khmelnik M.I. To the Question of the Source of Driving Force in the Searle's Generator. "Independent Authors Reports", publ. «DNA», printed in USA, Lulu Inc., ID 859217. Russia-Israel, 2007, issue 5 ISBN 978-1-4303-2444-7 (in Russian).

9. Khmelnik S.I. and Khmelnik M.I. Energy Processes in the Searle's Generator. "Independent Authors Reports", publ. «DNA», printed in USA, Lulu Inc., ID 1146081. Russia-Israel, 2007, issue 6, ISBN 978-1-4303-0843-0 (in Russian).

10. Tishina E.N. Magnetic cooling is already a reality. http://www.ndfeb.ru/articles/refreg.htm

11. Isaev S.I, Kozhinov I.A., et al The Heat Exchange Theory. M., «Vysshaya Shkola», 1979, 495 pages.

# Series: **Power Engineering**

**Boris E. Kapelovich, Solomon I. Khmelnik, David B. Kapelovich, Eugeny B. Benenson**

## Maintenance of Power Steam Turbine

### Abstract

The diagnostics system of the power steam turbine is offered. It can be executed also in the form of telediagnostic system. The system is presented on a site http://turbo.mic34.com/ System engineering can is ordered to authors.

### Contents

## 1. Problem

The modern level of power consumption demands constant increase of power of separate power blocks of heat stations and use of large amounts of fuel. In the conditions of the high market value of fuel, the preservation of the efficiency of power steam turbines (in abbreviated form - **PST**) is of paramount importance as even minor deviations of the operational parameters from the rated ones can cause excessive consumption of fuel and large material losses.

Presently, power stations have at their disposal plenty of precise measuring equipment and computers for processing the measurement results. However, it is traditionally accepted to determine only general power and economic parameters of the operation of the station, its blocks and shops. It is impossible to analyze the sources of power losses at this approach with the purpose of their elimination. To solve this problem, it is necessary to rely on the expertise of the maintenance personnel. But only in rare cases is it possible to determine quickly and accurately the cause of the abnormal condition of the **PST.**

Thus, the **PST** maintenance is characterized by the following factors:

- the general deviation of the power parameters from the rated ones of the *whole* PST is **not difficult** to detect,
- but **it is very difficult** to disclose the faulty component that causes this deviation;
- **it is very difficult** to determine quickly and accurately the economic losses and damage.

In order to preserve the deviations within the rated values in the process of the maintenance of **PSTs** of various power it is necessary to change technologically the scheme of the functional efficiency control. It is necessary to be able to determine losses in separate components of the **PST** technological scheme with the purpose of analysis of the cause of their origin and their elimination.

This problem can be solved by the Computer System of the power steam turbines efficiency analysis developed by us (in abbreviated form - **CS**). The CS

- formalizes the **search for the faulty component** of the PST,
- carries out **continuous monitoring** of the PST as a hole and each of its components,
- determines the **power losses** caused by the faulty equipment,
- gives the **personnel recommendations** on actions for elimination of the faults,
- assesses the **effect of the realizations** of these recommendations,
- collects the **statistics** of the components state changes for the optimal organization of repairs.

The system uses a new method of analyzing power losses in all components of the PST by the results of current measurements. The results of the analysis are later used for location of the faulty component and formation of recommendations. Following is a more detailed description of the system.

## 2. Computer System of power Steam Turbine Efficiency Analysis in the Process of its

The proposed system collects and processes information by the separate components and units of the PST with the purpose of determining local power losses. On the basis of the analysis of these losses, the system allows solving the following problems:

- Determination of the level and source of power losses in the steam cycle of the PST.
- Finding of the largest losses with the purpose of their prompt elimination.
- Analysis of the losses change in time. The investigation of the dynamics of various losses allows forecasting the further operation of the PST.
- The investigation of the local losses (especially in dynamics) allows correct planning of current and overall repairs, organizing rational preparation for these repairs.
- Analysis of local losses allows increasing interrepair time or, on the contrary, to make unplanned repairs to avoid emergency situations.

On the basis of the developed algorithms and mathematical models, it is possible to determine losses caused by the following factors:

- decrease of internal relative efficiency in the high pressure cylinder;
- decrease of internal relative efficiency in the medium pressure cylinder;
- clogging of the intermediate superheater;
- deterioration of vacuum in the condenser;
- overcooling of the condensate in the condenser;
- failure of performance of each regenerative heater of the feeding water.

All these losses can be calculated over an hour, day, month and year of operation and expressed in terms of natural fuel, equivalent fuel and money. The amount of losses is shown on the display beside the corresponding component of the technological scheme. These parameters are then compared with the rated values specified by the manufacturer or with the results of heat tests.

The calculation of the local losses of the **PST** can be made by two methods:

- determination of daily fuel losses; measurements and calculations are made automatically during the day with the subsequent summation of all fuel losses;
- determination by one rated load.

The first method is more laborious and requires a great number of initial data received during a complex test of the **PST** in a wide range of loads. The second method does not require testing of the equipment and allows confining to the data of the manufacturer. However, the second method also allows making <u>fairly good assessment of the equipment condition</u>.

The following variants of the system are possible:
- analysis with the fixed power of the equipment;
- analysis with the variable power of the equipment;
- analysis with the formation of an archive and statistics.

The analysis is possible:
- with manual data input;
- with automatic data input.

## 3. Effect of the System Realization

The economic effect of the system realization depends on the turbine power, the heat scheme parameters, the operation mode and other factors characterizing the operation of each PST. The received measurements of a **228 MW** operational Parson turbine were used for the evaluation of its characteristics (see below). Even with rather moderate parameter deviations from their rated values, the total overconsumption of equivalent fuel is **17.31 tons** per day. If the faults are eliminated promptly and only basic losses are reduced, the fuel economy will be **2,854.3 tons** a year, which in money terms will amount to **$285,430** a year at the fuel price of **$100** per ton. The implementation of the system makes it possible to save up to **$1000 per 1 MW** of the established power a year (depending on the fuel cost, the unit power and type).

# 4. About the System Implementation

The proposed system can find wide application in all power blocks of any power stations, as well as in large ship power installations. The system can be used in the turbine equipment of heat stations. But the greatest effect is reached in large power blocks condensing.

The implementation of the system does not require large capital spendings as it is based on the use of regular control devices of power stations. The basic expenditures fall to designing and introduction CS.

When first-priority objects for the system implementation are chosen, it is necessary to take into account the fact that its efficiency depends on the unit power and fuel cost. Therefore, large power blocks condensing, and stations using expensive fuel should be preferred.

The developed algorithm of determination of local losses is based on the project calculations of the manufacturer, the results of the heat tests of the PST (if available) and **general principles of the thermodynamics and theory of steam turbine devices**. Therefore, for the system realization on the chosen equipment, the following information should be provided:

1) the power block heat scheme with the designation of all rated parameters;
2) the results of the latest heat tests of the unit at the power station;
3) the following characteristics of the condenser:
   - $P_k = f(D_k; T_b)$, $N_e = f(P_k)$, where
   - $P_k$ is pressure in the condenser;
   - $D_k$ is steam consumption in the condenser;
   - $T_b$ is cooling water temperature at the condenser input;
   - $N_e$ is the steam turbine power;
4) specified power dependences of specific fuel consumption:
   - $B = f(N_e)$, where
   - $B$ is fuel flow,
   - $N_e$ is the steam turbine power.

# 5. The System Implementation Variants

Any variant of the system implementation presupposes the input of the results of current measurements of the PST equipment parameter into the system (at certain intervals).

The system

- calculates the local and total losses in the PST equipment over the specified period,
- assesses the economic effect after the elimination of the losses,
- gives recommendations on the optimization of the equipment repairs.

The variants of the system implementation differ only in methods of measured parameters entering. However, these methods affect essentially the cost of the system and its operation. We propose the following options.

### A. Automatic System.

The system software is installed in the power station computer net. The regular control devices should have terminals on the control board. Their accuracy is adequate for the assessment of losses. All the devices should have outlets for connecting to the computer net. The automatic system allows fixing the measured parameters automatically at specified intervals and provides the calculated deviations and losses at the same intervals.

### B. Non-automatic System

This variant differs from the previous one in that it does not require additional electronic equipment for control devices. There is only one condition – the availability of software and a computer. The data are entered manually from the daily sheet by one rated load.

### C. Telediagnostic System

This option is the simplest for the user. The maintenance is organized in the following way. The PST duty personnel ("*the user*") sends the data from the daily sheet to the *server* by e-mail. The user sends certain data by one rated load only. In reply, the server returns the user

- the local and total losses of the PST over the specified period and in dynamics,
- recommendations on the repair and maintenance,
- the assessment of the economic effect after the realization of these recommendations.

The proposed diagnostic system solves these problems on the basis of the local losses analysis. To make this analysis the daily sheet data are quite sufficient.

# 6. Example

Below are the results of the computer system operation for a Parson PST with the following parameters: $\mathbf{Ne}$ = 228 MW; $\mathbf{P_0}$ = 14 MPa; $\mathbf{t_0/t_{01}}$ = 538/538°C. The deviations from the nominal values are as follows:

| | |
|---|---|
| steam temperature at the outlet of the high pressure cylinder $(\mathbf{T_1})$ | +2°C; |
| steam temperature at the outlet of the medium pressure cylinder $(\mathbf{T_2})$ | +3°C; |
| pressure in the intermediate superheater | 11%; |
| steam temperature in the condenser $(\mathbf{t_k})$ | +3°C; |
| overcooling of the condensate $(\mathbf{t_k^1})$ | -1°C. |
| feeding water temperature after high-pressure heater 5 $(\mathbf{t_5^1})$ | -3°C; |
| condensate temperature after low-pressure heater 3 $(\mathbf{t_3^1})$ | -2°C; |
| condensate temperature after low-pressure heater 2 $(\mathbf{t_2^1})$ | -2°C. |

The calculation results are shown in the PST scheme on the computer display while the following terms are used:

- **losses** – fuel losses due to the faulty equipment measured in tons of equivalent fuel per day;
- **component** – equipment that is part of the PST technological scheme;
- **parameter** – measured value;
- **parameter deviation** – the result of subtraction of the parameter measured value from the rated one.

On plan PST each component is accompanied by an information element of the following view:

In an information element are allocated

- **field 1** – component number;
- **field 2** – losses in the component (in tons of equivalent fuel per day);
- **field 3** – the field for input of the parameter deviation.

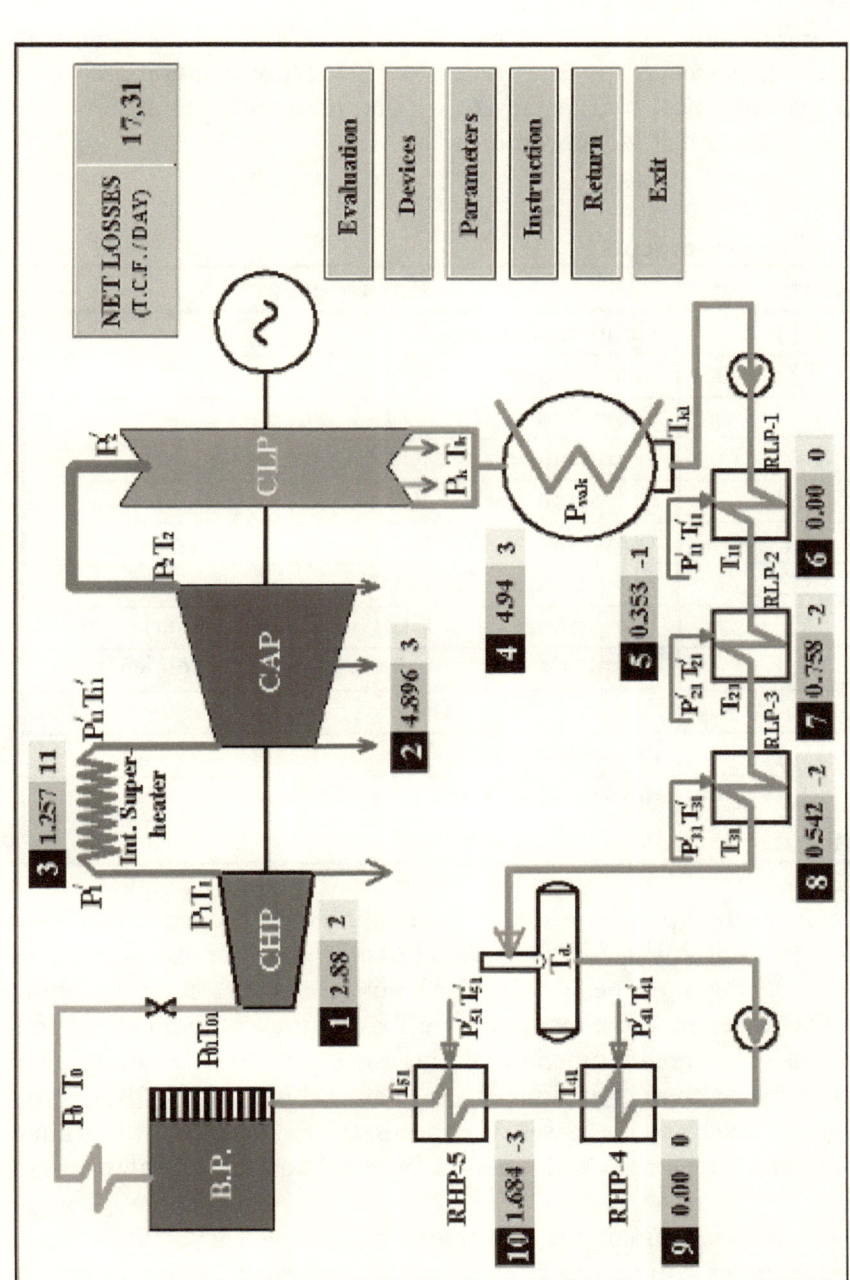

In addition, the operator can see **information tables**. The table "**Losses**" shows the losses in the components and the corresponding parameter deviations. For deviations, the physical limits of measurement are indicated, as well as the discreteness of their representation and their dependence on measured parameters. The measured parameters are indicated in the table "**Measurements**".

*Table* "**Measurements**"

| Denotation | Parameter |
|---|---|
| T11 | Temperature after RLP-1 |
| T21 | Temperature after RLP-2 |
| T31 | Temperature after RLP-3 |
| T41 | Temperature after RHP-4 |
| T51 | Temperature after RHP-5 |
| T1 | Temperature after CHP |
| T2 | Temperature after CAP |
| Tk | Temperature after CLP (in the condenser) |
| Tk1 | Temperature after (in the collection of a condensate) |
| P1 | Pressure after CHP |
| P11 | Pressure before CAP |
| C=2.57at | Resistance of the intermediate over heater |
| normX | Norm of parameter X |

The measured parameters are entered into the system while the deviation value is calculated. It is also possible to enter the parameter deviations directly into the system. The value of the parameter deviation affects the losses in a component. The loss in the component can be determined as a certain function of the corresponding deviation. The determination of these functions is the main subject of the developed program. To calculate the losses, it is necessary to determine the values of all parameter deviations and press the button "**Losses Calculation**".

When the measurements are entered automatically, the system program operates continuously, releasing the values of the losses in the on-line mode.

*Table* "**Losses**"

| N | Delta | Min | Max | Deviation of parameter | Unit | Losses (ton cond. Fuel/day) |
|---|---|---|---|---|---|---|
| 1 | 1 | 1 | 10 | dT=normT1-T1 | °C | **2.88** from deterioration of efficiency in CHP |
| 2 | 1 | 1 | 10 | dT=normT2-T2 | °C | **4.896** from deterioration of efficiency in CAP |
| 3 | 1 | 11 | 13 | dP=[(P11-P1)/C-1]*100 | °C | **1.257** from pollution of the superheated |
| 4 | 1 | 1 | 10 | dTk=normTk-Tk | °C | **4.94** from deterioration of empty cpace |
| 5 | 1 | -5 | -1 | dTp=Tk1-Tk | °C | **0.353** from overcooling a condensate |
| 6 | 1 | -5 | -1 | dT11=T11-T11 | °C | **0.0** from underheating in RLP-1 |
| 7 | 1 | -5 | -1 | dT21=T21-T21 | °C | **0.758** from underheating in RLP-2 |
| 8 | 1 | -5 | -1 | dT31=T31-T31 | °C | **0.542** from underheating in RLP-3 |
| 9 | 1 | -5 | -1 | dT41=T41-T41 | °C | **0.0** from underheating in RHP-4 |
| 10 | 1 | -5 | -1 | dT51=T51-T51 | °C | **1.684** from underheating in RHP-5 |

# References

1. Kapelovich B.E., Khmelnik S.I., Levshin A.U., Kapelovich D.B., Benenson E.B. Maintenance of Power Steam Turbine. «The Papers of Independent Authors», publ. «DNA», Russia-Israel, 2005, vol. 2, p. 181, printed in USA, Lulu Inc. №172756, ISBN 978-1-4116-5956-2 (in Russian).

2. Kapelovich D.B., Kapelovich B.E., Khmelnik S.I. Maintenance of Power Steam Turbine. «Energetika ta elektrifikazia», Ukraine, 2007, №5 (285), p. 30

## Series: **Unaccounted**

**Genady G. Filipenko**
# About decoding of "The Stormer Effect"

## Abstract

The phenomenon stated by C. Stormer is discussed – "The Echo of Short Waves, Which Comes Back in Many Seconds after the Main Signal". The hypothesis about a content radio echo is offered.

The phenomenon is described by C. Stormer in his work 'The Problem of Aurora Borealis' in the chapter entitled 'The Echo of Short Waves, Which Comes Back in Many Seconds After The Main Signal' [1]. In 1928 the radio- engineer Jorgen Hals from Bigder near Oslo informed C. Stormer about an odd radio echo received 3 seconds after the cessation of the main signal; besides, an ordinary echo encircling the Earth within 1/7 of a second was received. In July Prof. Stormer spoke to Dr. Van-der-Paul in Andhoven and they decided to carry out experiments in autumn and send telegraphic signals in the form of undamped waves every 20 seconds three dashes one after the other. On 11 October 1928 between 15.30 and 16.00, C. Stormer heard an echo 'beyond any doubt'; the signals lasted for 1,5- 2 seconds on undamped waves 31,4 meters long. Stormer and Hals recorded the intervals between the main signal and the mysterious echo:

1) 15, 9, 4, 8, 13, 8, 12, 10, 9, 5, 8, 7, 6
2) 12, 14, 14, 12, 8
3) 12, 5, 8
4) 12, 8, 5, 14, 14, 15, 12, 7, 5, 5, 13, 8, 8, 8, 13, 9, 10, 7, 14, 6, 9, 5
5) 9

Atmospheric disturbances were insignificant at that time. The frequency of echoes was equal to that of the main signal. C. Stormer explained the nature of echoes by reflection of radio waves from layers of particles ionized by the Sun. But! The Professor of the Stenford Electrotechnical University R. Bracewell suggested possibility of informational communication through space probes between more or less developed civilisations in space. From that point of view the information about decoding of Stormer series can be found in following journals [2, 3, 4, 5].

The author of this work offers the following decoding: let the numbers in the series be replaced for chemical symbols of elements with corresponding nuclear charges:

1) P, F, Be, O, Al, O, Mg, Ne, F, B, O, N, C
2) Mg, Si, Si, Mg, O
3) Mg, B, O
4) Mg, O, B, Si, Si, P, Mg, N, B, B, Al, O, O, O, Al, F, Ne, N, Si, C, F, B
5) F

It is easy to see that the second series is repeated at the beginning of the forth series with the only difference that in the forth series silicon is alloyed with boron and phosphorus, i.e. 'p-n transition' of a diode is created. The third series describes receipt of pure boron through action on boron anhydrite by magnesium: $B_2O_3 + Mg \longrightarrow B+...$

The author of the above hypothesis wrote his degree paper on silicon carbide light-emitting diode, that is why the ending of the forth series is the most simple- it is a modern light-emitting diode. Silicon carbide is alloyed with nitrogen and boron with 'some participation' of fluorine. Approximately the same way diamond is alloyed with participation of fluorine in laboratories of 'other civilisations', as can be seen at the ending of the first series. In the middle of the forth series corundum, the base of ruby, is also alloyed with boron, nitrogen and fluorine. In the fifth series simply fluorine is educed as a useful but very aggressive gas. Inert neon seems to divide optoelectronic devices.

In conclusion, some repeated applications should be noticed:
- fluorine favours in a way either diffusion of boron;
- fluorine favours in a way either electronic processes in forbidden zones of diamond, silicon carbamide;
- for some reason magnesium contacts are used.

In 1928 semi-conductor devices were not in use on Earth, that proves an extraterrestrial origin of the set forth above information.

# References

1. K. Stormer. Problem of Streamers, publ. "Gostechizdat", 1933, Russia (in Russian).
2. F.U. Zigel. The Stormer Paradox, magazine "Smena", № 2, 1966, Russia (in Russian).
3. Ronald N. Bracewell. The opening message from an extraterrestrial probe, "Astronautics and Aeronautics", №5, USA, 1973.
4. Anthology of enigmatic Events, magazine "Техника молодежи", №4, 1974, Moskow, Russia (in Russian).
5. Anthology of enigmatic Events, magazine "Техника молодежи", №5, 1977, Moskow, Russia (in Russian).
6. H. Filipenko, The Stormer Effect, http://www.belarus.net/discovery/filipenko/fil2.htm, 1998.
7. Filipenko G.G. About decoding of "The Stormer Effect". «The Papers of independent Authors», publ. «DNA», Russia-Israel, 2005, vol. 1, p. 37, printed in USA, Lulu Inc. #124173 (in Russian).

# Authors

**Benenson Evgeny B.,** *Israel.*
benenson@012.net.il
St.-Petersburg was born in 1936. Has ended North-western Polytechnical institute and the Ivanovo Power Institute. Was engaged in an electrospecial equipment of planes and engines, дефектоскопией metal of the power equipment of power stations, technological and design development.

**Gelfand Alexander M.,** *Russia.*
Delfin.Informatika@delin.ru
The assistant to the general director of Open Society "Institute Energosetproekt", the leader of research-and-production company "Delphin-Informatika". Area of interests: the system analysis; methods of the analysis and data processing; the integrated control systems. It is published more than 50 clauses.

**Filipenko Genady G.,** *Belarus.*
hfil@aport2000.ru
Was born in 1948 and has grown in Belarus. Lives in a of Grodno. After service in army has finished the Leningrad Electrotechnical Institute on a speciality «Semiconductors and dielectrics». Was engaged in cultivation of monocrystals iron-itriy and calcium-bismuth-vanadic of pomegranates; powder metallurgy; a crystal-chemistry, a crystal-physics.

**Karpov Mihail A.,** *Russia.*
karpovm@mail.ru
Was born in 1959. In 1981 has ended radiophysical faculty of Gorki State University on a speciality «Radiophysics and electronics». Area of interests - physics of elementary particles and cosmology. I am married and I have two sons.


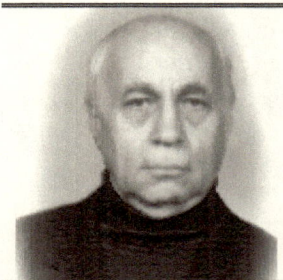

**Kapelovich Boris E.,** *Israel.*
The professor of the Ivanovo Power University, the author of 80 articles, inventions and books, including books «Maintenance of steam-turbine installations», 1975 and 1985; the textbook «Maintenance and repair of steam-turbine installations», 1988.

**Kapelovich David B.,** *Ukraine.*
Candidate of engineerings sciences, senior research scientific employee. Director of the State scientific-research Institute "Teploenergetika", Ministry for the Power Generating Industry of Ukraine. The author of 12 articles and 15 copyright certificates.

**Khmelnik Mihail I.,** *Israel.*
solik@netvision.net.il

The doctor of physical and mathematical sciences. Scientific interests-hydrodynamics, the theory of a filtration, a current in gases, mathematics. Has about 120 scientific articles. Has prepared a number of candidates and doctors of sciences. Many years worked the senior lecturer, and then the professor of the Moscow state university of press. Many years were the scientific secretary of a seminar on hydrodynamics at Institute of problems of mechanics AH (the USSR, and then the Russian Federation), the scientific secretary of section of physics of the Moscow society of testers of the nature at the Moscow State University. The honourable professor of the Kirghiz state university of construction, transport and architecture.

**Khmelnik Solomon I.,** *Israel.*
solik@netvision.net.il

Ph.D. in technical sciences. Author of about 150 patents, inventions, articles, books. Scientific interests – electrical engineering, power engineering, mathematics, computers.

www.ingramcontent.com/pod-product-compliance
Lightning Source LLC
Chambersburg PA
CBHW022025170526
45157CB00003B/1353